Earth's Unseen Inhabitants

Fast Movers, and Quantum Communication

Earth's Unseen Inhabitants

Fast Movers, and Quantum Communication

By Wayne Lawrence, Bill Spicer, Lily Nova, and Margie Kay

Printed in the United States of America

ISBN: 978-1-969800-07-8

Library of Congress Control Number applied for

Cover design by Wayne Lawrence

Photographs copyright by the authors

Publisher

Un-X Media
PO Box 1166
Independence, MO 64051
unxmedia.com

EARTH'S
UNSEEN
INHABITANTS
Fast Movers and Quantum Communication
Margie Kay, Wayne Lawrence, Bill Spicer, Lily Nova

Table of Contents

FOREWORD

In the first edition of *The Fast Movers*, co-authors Margie Kay, Bill Spicer and Wayne Lawrence showed what we have been observing and cataloguing for several years as we have investigated different cases of UFOs/UAPs-- ending in 2019.

This book is an update of new evidence of Fast Movers, and covers the period from early 2020 until June 2023, a total of about 3 years. We will spend more time discussing not only the Quantum UFO Observation Technique (QUOT) to observe fast movers and stationary craft, but also how forms of non-local Quantum Communication might be explained as a framework for communication in non-traditional ways with "unseen inhabitants." This has been experienced by the authors of this book.

Scientific study of the quantum universe, a.k.a. quantum hologram, is a burgeoning field with many expressions of communication from intuition, mental telepathy, clairvoyance, remote viewing, ESP, precognition, and others. It appears to be a method of contact for many similar phenomena. The experience of the authors with non-local quantum communication will be part of this edition and is framed by theory and experimental evidence of many scientists and researchers.

One of the driving forces behind the writing of this second volume since "*The Fast Movers*" is the realization that all three of the authors had experienced elements of quantum communication. As we move forward in our investigations we realize that "*...genuine psi* suggests that what science knows about the nature of the universe is seriously incomplete ...*"

Research has shown that 24% of study participants who had experienced contact with Non-Human Intelligence (NHI) had "deliberately called an NHI." Additionally, the study also showed that 78% of respondents reported communications with NHI. It is becoming very clear that humans can "...exchange information and affect change at a distance." (Beyond UFOs, pp. 51, 84).

We have found these statements to be accurate in our experience as well.

"As the twenty-first century dawns, astounding new visions of reality are stirring."

(See: ***BEYOND UFOs***, *The Science of Consciousness and Contact with Non-Human Intelligence, Vol. 1, Ed. Rey Hernandez, JD, MCP, Dr. John Klimo, PhD, Dr. Rudy Schild, PhD, 2018, pp. 51, 84, 368-390.)*

(See also: ***THE CONSCIOUS UNIVERSE***, *The Scientific Truth of Psychic Phenomena, Radin, D., 1997, pp. 325, 340)*

Section 1 by Wayne Lawrence

A Brief Introduction to The Quantum Concept

Our first introduction to the term "quantum" may have been as we studied basic science in our secondary educations. It would have been in connection with "**quantum** mechanics."

The word was introduced in physics directly from Latin by Max Planck, 1900, on the notion of the "minimum amount of a quantity which can exist;" later reinforced by Albert Einstein, 1905. Later it described the new understandings of the workings of the atom as "**quantum theory**," 1912, and its relationship to using that knowledge practically in "**quantum mechanics**", 1922. Today, we often use a variation of it by associating it with "**quantum leap**" to mean a breakthrough in understanding. In a real sense, which can also describe what we are beginning to experience as we slowly adapt to new understandings of our universe, and how we interact with that universe.

The universe is regarded by many scientists as a "Quantum Hologram," and it can store and retrieve the content of life existence, (emotional content). The Quantum Hologram is sometimes called "nature's mind." Quantum non-locality (being everywhere) is the basis of perception which means that (human) perception is available everywhere at any time. (Hernandez, etc…, *Beyond UFOs,* pp 368-387). Both quantum information and space-time information are used in the act of perception by organisms.

For example, after a decade of research at the Princeton Engineering Anomalies Research Laboratory, it has been shown that humans can exchange information without the use of their normal sensory systems and cause change in other people and physical systems at a distance (telepathy, remote viewing, psychokinesis), using mental concentration alone. (Hernandez, etc…*Beyond UFOs*, pp. 87, 385 op. cit.)

The impact of the foregoing statement is monumental and begins to shatter our Newtonian and Cartesian world views which were the basis of the first scientific revolution, centuries before Quantum Physics entered our understanding. We owe a debt of gratitude nonetheless to scientific greats such as Copernicus (1543), Galileo (1610), Descartes (1637), Newton (1687) and many others who helped to guide our understanding as a species to the world that we have known. Their discoveries helped usher in the scientific revolution and take us out of the Dark Ages. Yet we are facing some monumental resistance from the same individuals who embrace the scientific method.

Quantum communication begins to provide an understanding of what underlies some of the Contact Experiences such as: Contact with NHI, Out-of-Body, Near-Death, and Remote Viewing experiences, and PSI Phenomena such as: telepathy, precognition, clairvoyance, and psychokinesis.

Quantum Communication, as it is used by the authors of this work, is described as intentional communication with Non-Human Intelligence (NHI) usually by mental means, and includes intentional contact from either direction:

NHI to Human, Human to NHI or Human to Phenomena. For additional information regarding the topic of Quantum Communication in relation to the cases that we are highlighting, please refer to the beginning pages of Chapter 2—**QUANTUM COMMUNICATION**, CONSCIOUS INTENTION.

WHY ARE WE TALKING ABOUT ALL OF THIS?

Because the evidence of Fast Movers, the realization of conscious communication in interactions with NHI, & experiences with remote viewing, are all indicative of a pervasive presence of other NHI on Earth.

These are Earth's Unseen Inhabitants. Many researchers believe they have been here a long time…

> "*It hasn't just visited; it has been here a long time, and it is still here.*"
> --Dr. Garry Nolan, PhD, in: "Secret UFO Briefing," by John O'Sullivan, *Irish Star*, Jan. 1, 2024.

"I happen to be privileged enough to be in on the fact that we have been visited on this planet, and the UFO phenomenon is real."

Edgar Mitchell

Chapter 1

The FAST MOVERS, Updated

Additional Cases from 2020 through 2023

By Wayne Lawrence

The first volume that we collaborated on is entitled "The Fast Movers" and covered a period ending in 2019. Since that time, 20% of the cases that I have investigated for MUFON have been Fast Movers. There undoubtedly are many more but as with unidentified flying objects generally, most go unreported.

It has been my experience that random videos taken over bodies of water have always captured Fast Movers. Perhaps this is because they are just so numerous that they have almost always been recorded on video, and occasionally the lucky photo. There may also be a quantum communication linkage to the phenomenon. I say this because Bill Spicer has taken a more active role in requesting that UFOs show themselves, and quite often they do after he has sent out a mental message. He will cover more on his experience in his section of the book.

This first chapter is about…

Providing an update with additional "Fast Movers" cases that I have investigated and that I regard as some of the most interesting. We find "Fast Movers" in many ways, and some represent a more passive approach. When cases of sightings were analyzed, sometimes fast movers are discovered—not by the witness—but during the investigation of their cases. I have found that when witnesses take videos of what they think is a UFO, often Fast Movers are caught…and they are not even the object of the video. **Fast Movers usually are <u>not noticed</u>** because they travel so rapidly. Most of the time they travel through the screen in less than a second. In this chapter, I will show some of the more interesting images from 9 cases. I have investigated 21 cases that included Fast Movers these last three years.

FOUR GLOWING CRAFT

Case # 108984
Date: Sat. May 16, 2020
Time: 9:02 pm
Place: Liberty, MO

This image-- # 2237-- was taken 43 seconds after image 2236, and 48 seconds prior to 2238.

On Saturday May 16, 2020, this case was witnessed by journalist Christian Vedder who was photographing his classic car (1964 Ford Galaxie) on the square in Liberty, MO. (*The same model car that was used in filming Men in Black III*). He took 8 photos.

Later when he looked at the images, he realized that there were 4 to 5 objects that he had not seen with his eyes in the next-to-the-last photo. They obviously had been just caught by the camera in a fraction of a second. He did not notice them before or afterward. If they had lingered in the sky, they most likely would have been seen.

After our initial contact, Christian was urged to return to the location of the sighting and open the SkEye App, then take a screenshot of the view of the sky and associated stars and constellations. Knowing the closest star, and after consulting star charts at that time of day, one can know the angle of elevation of the star at that time of day. This is one way to know the angle of observation during the sighting. (The image below was taken during daylight hours six days later).

The angle of observation was 23.5 degrees to the largest and brightest of the objects. Since the cloud ceiling was at 1,363 ft, and the final object at the far right appears to be coming out of the clouds, the cloud base is a good indicator of the altitude of the craft. The base distance from the witness to directly beneath the objects was .59 miles. This allows us to calculate the size of the brightest craft to be about 30.5 feet. (Calculations on next two pages).

I asked Mr. Vedder to center the app on the area of the craft in his photo. The afternoon that he took it the constellation ***Virgo*** *was there. I checked the star charts for that particular star that is near the hand. That is* ***Spica****. 2.5 degrees below Spica is where the craft was…or 23.5 degrees. He took the image on May 22 at 7:48 pm. I looked up the star, for that day and time, and Spica was at 26 degrees. So, the craft was at 23.5 degrees!*

How to understand the data in the calculations <u>on the next page</u>

Every observation of anything in the sky is a right triangle. That means that the bottom angle under the object is at 90 degrees, and that the other two angles of the triangle vary.

- The distance represented by **a°** is the distance from the ground up to the object.
- The letter "**a**" found on the bottom of the triangle is the distance from the witness on the ground to a point directly underneath that object.
- The letter "**c**" is used for the hypotenuse (although the "c" is not labeled in the diagram) and represents the direct line-of-sight view of the witness to the object in the sky.

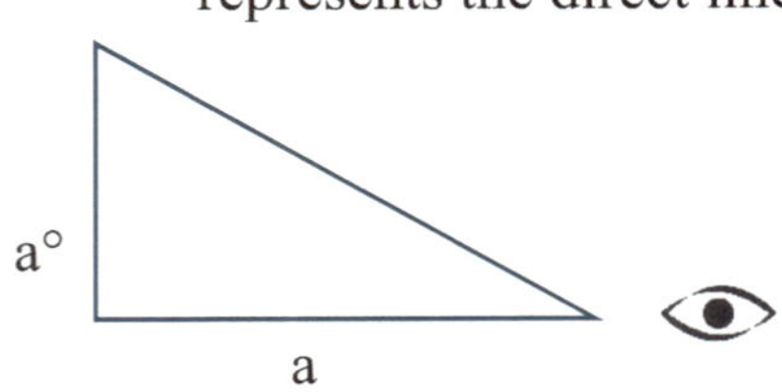

CALCULATIONS: Case # 108984

Measuring the size of a craft

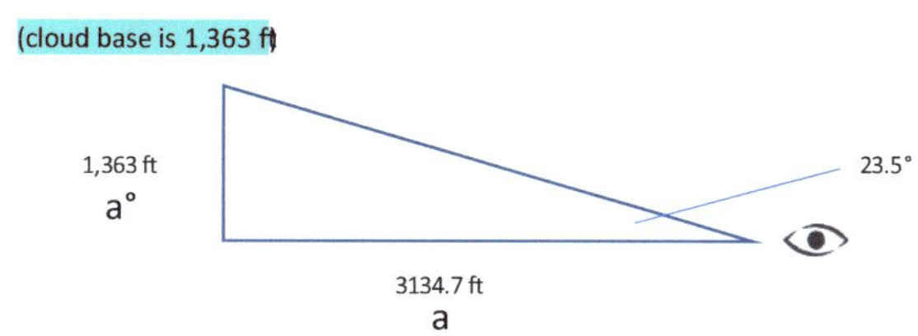

$$\frac{a^\circ}{a} = \text{tangent } \Theta$$

$$\frac{1{,}363 \text{ ft}}{a} = .4348 \qquad \textit{(tangent of 23.5°)}$$

$$\frac{1{,}363}{.4348} = a$$

a = 3,134.7 ft (ground distance)

Now… I measured the craft on my computer screen:

98 mm = screen height (from craft to horizon) = 1,363 ft

2.5 mm = length of craft on screen

$$\frac{1{,}363 \text{ ft}}{98 \text{ mm}} = \textbf{13.9 feet/mm}$$ **for screen measurement**

Size of Craft:

2.5 mm = length of craft on screen

2.5 mm x 13.9 ft/mm = **30.58 ft**

Enhancements of the brightest craft:

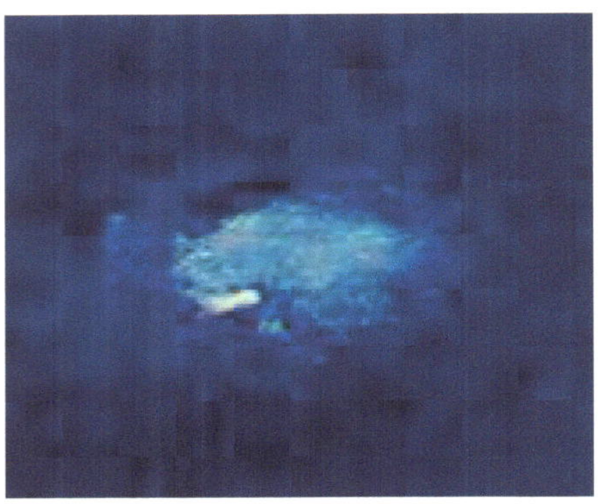

Luminance study

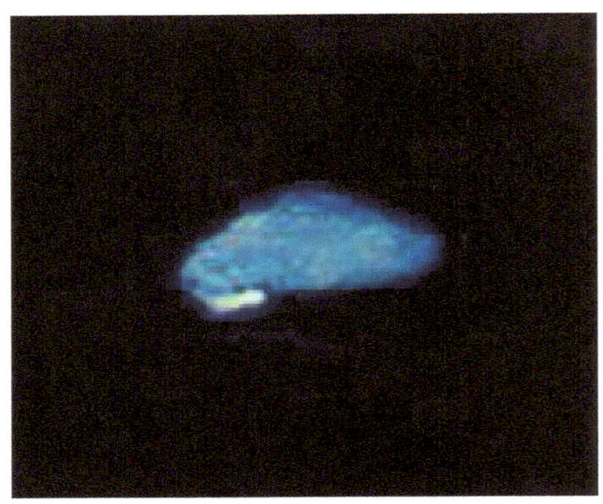

Additional enhancement

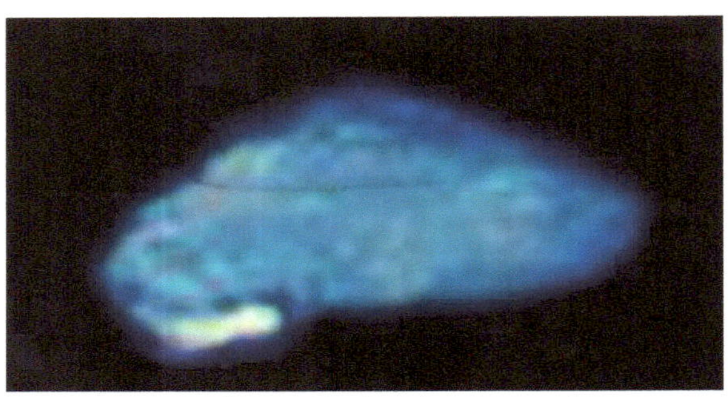

Line accentuated separating lower and upper sections

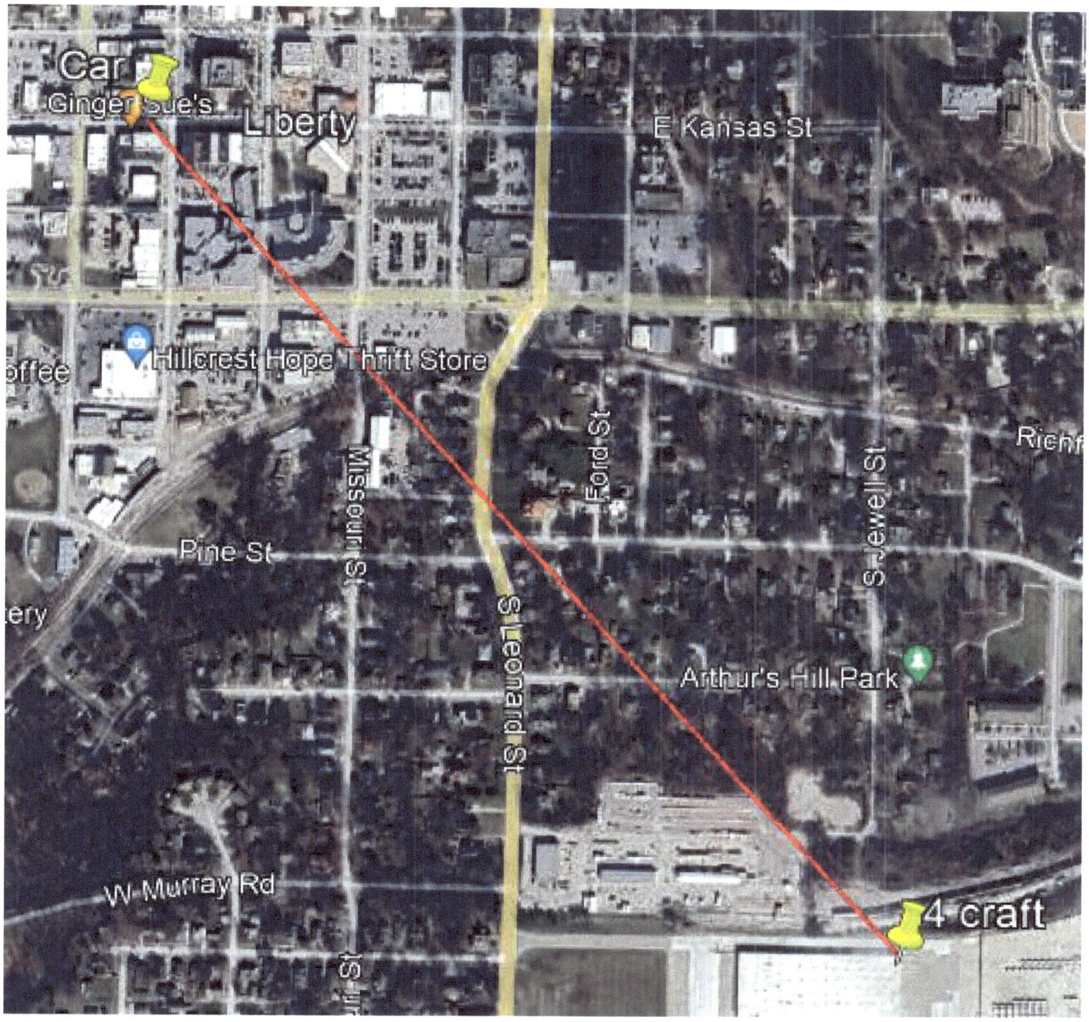

CONCLUSION:
The four craft appear to have come down out of the clouds, though the argument could be made that they were entering the clouds as well. Christian stated that he did not know whether they were coming or going. Regardless, they did not linger and were apparently only present long enough to be briefly captured by the camera.

- Known aerial vehicles would have been seen traveling through the sky-- but these were not seen. Likely they were traveling rapidly.
- There was no signature of sound from the objects that were heard.
- No propulsion system was evident.

It appears that Vedder captured at least four Fast Movers in the next to last photo. They were approximately 3,135 ft distant and at an altitude of 1,363 ft. The brightest one measured roughly 30.5 feet long.

NOTE: There is a Remote Viewing connection with this case. For more detail, see page 54.

Fast Movers at Lowe's

Case # 109229
Date: Fri. May 29, 2020
Time: 3:22 pm
Place: Liberty, MO

On Friday May 29, 2020 a witness saw what he thought to be an unusual object that he could not identify. He took a couple videos of it.

The object happened to be a jet, but unbeknownst to the witness, eight Fast Movers were also caught in the video.

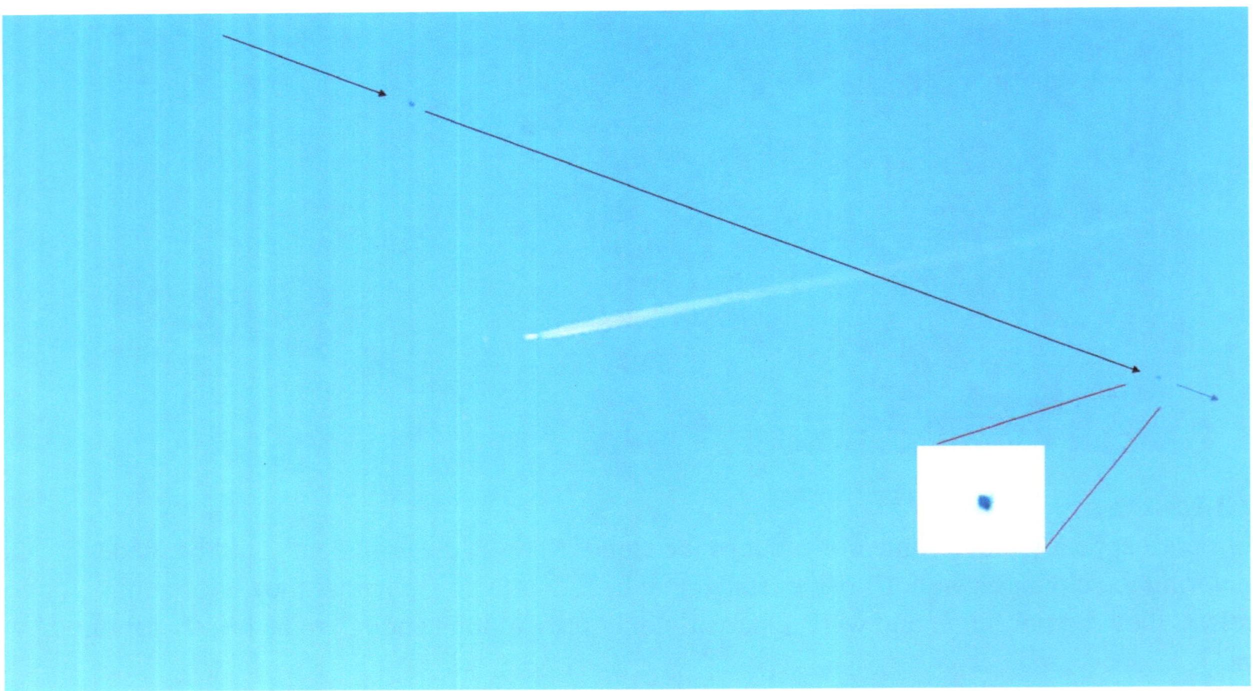

Merged frames 4:27 and 5:01—Object moved through the scene in .16 seconds

Two of the examples of the Fast Movers that traveled through the video are presented in this report. This will provide the basic idea of what they look like, and how rapidly they truly are.

Since one of the Fast Movers did run through the contrail, we can calculate distance, and speed from the aircraft's altitude.

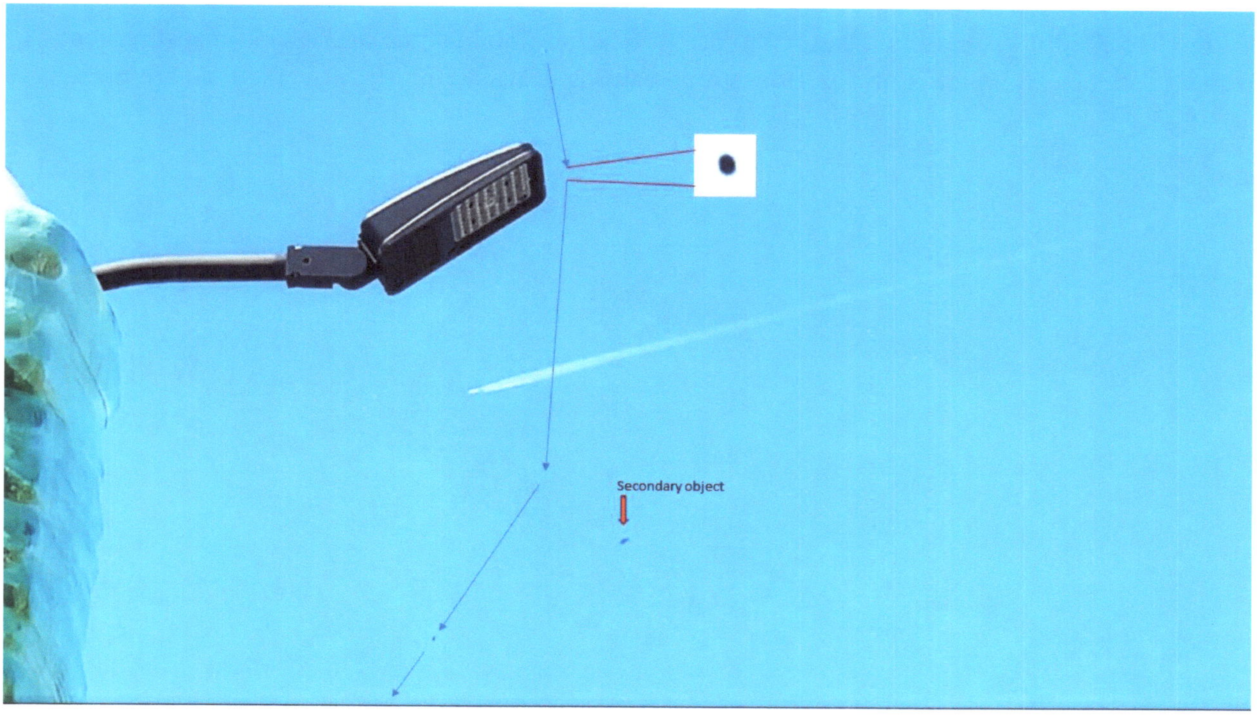

Frames 34:09 to 34:16

Object moved through contrail and screen in .27 seconds (approx. 14,148 mph).

Calculation of the size of the object recorded by the witness

Angle of elevation of the object from the horizon: 25 degrees

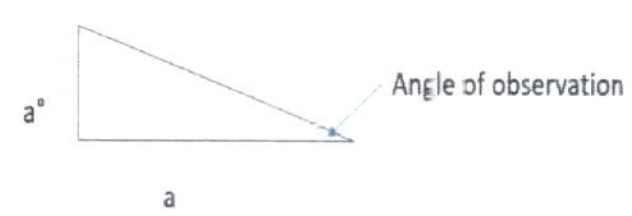

Note: 26,000 ft is the jet's approximate altitude.

Land distance miles:

$\frac{a^\circ}{a}$ = tangent Θ

$\frac{26{,}000 \text{ ft}}{a}$ = ***tangent*** of 25 degrees = .466 a = 55,793 ft, or 10.6 miles away

---

Line of sight miles:

$\frac{a^\circ}{c}$ = sine Θ

$\frac{26{,}000 \text{ ft}}{c}$ = ***sine*** of 25 degrees = .4226

c = 61,524 ft, or **11.6 miles** line of sight distance

Length of craft:

Ratio of aircraft size compared to altitude = .0045

(obtained by screen measurement of the craft divided by the distance from the horizon to craft).

Multiply 26,000 ft x .0045 = **117 ft long,** approximately the same size as an Airbus 220, 319 or 320. So the sighting by the witness is confirmed as a large jet aircraft. The importance is that the angle of observation is established at 25 degrees which assisted in the calculations of the speed of the unknown fast mover .

On Sunday June 21, 2020, I visited the location of case 109229 to obtain some angles and measurements. I took some random videos and photos. In the videos, one can see at least 16 Fast Movers. Two are presented here…

Frames 5:15 to 5:17 object changing shape in .1 seconds.

Frames 14:27 to 15:07 then disappears in **.37** seconds

In another captured set of images from the video that was shot, one of the objects calculated to be approximately the size of a basketball. It was found **at an altitude of 3,863 feet** (cloud base), traveling at about **4,808 mph.** Then it split into two. The image at right is the result.

A UFO Hunt

Fast Movers Captured on Video

Date: Friday June 30, 2020
Time: several hours
Place: Three Lee's Summit, MO lakes

Small Black Sphere over **Longview Lake** viewing West at 280°

I decided to go on a hunt for UAPs since I had been seeing so many of them.

There was a great deal of resolve and intention as I went to Longview Lake, Prairie Lee Lake, and Lake Jacomo to film whatever I could find. Four of the 23 fast movers filmed that day are depicted in this report.

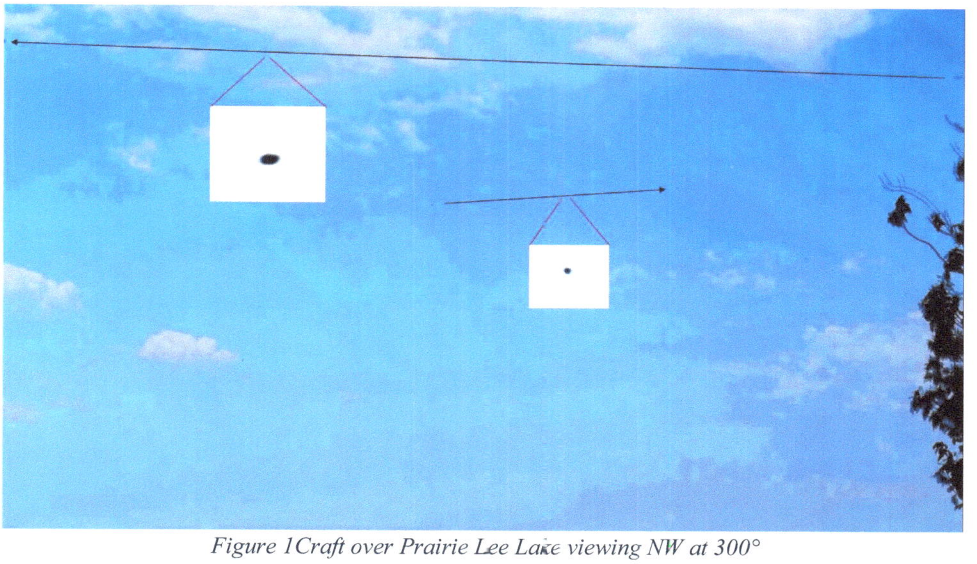

Figure 1Craft over Prairie Lee Lake viewing NW at 300°

The larger craft traveled from frames 3:24 to 4:03 in 10/30th of a second, **or about 1,415 mph,** a distance of 8 degrees azimuth from 300 to 292 degrees, covering .13 land miles.

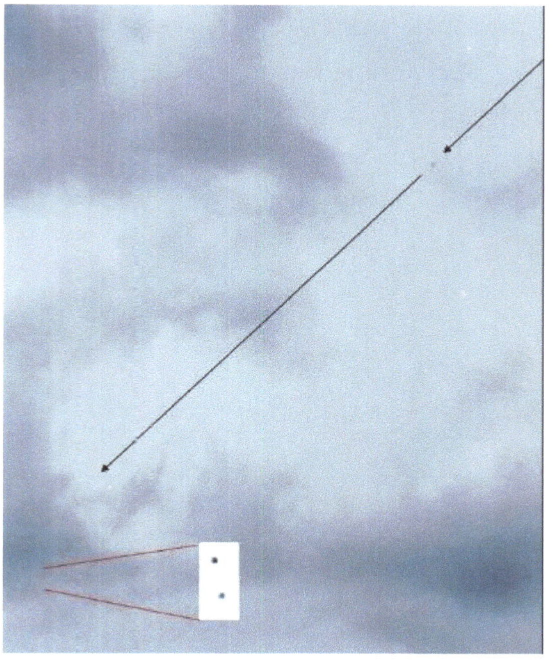

South Lake Jacomo

Facing North

The object covered .255 miles from bottom to top of screen in .73 seconds, traveling at approximately **1,257 mph**. Depicted above are frames 13:27 to 14:18. The size of the object calculates to **8.5 feet** in diameter, with a total of 226 square feet of surface area, and 321 cubic feet in volume.

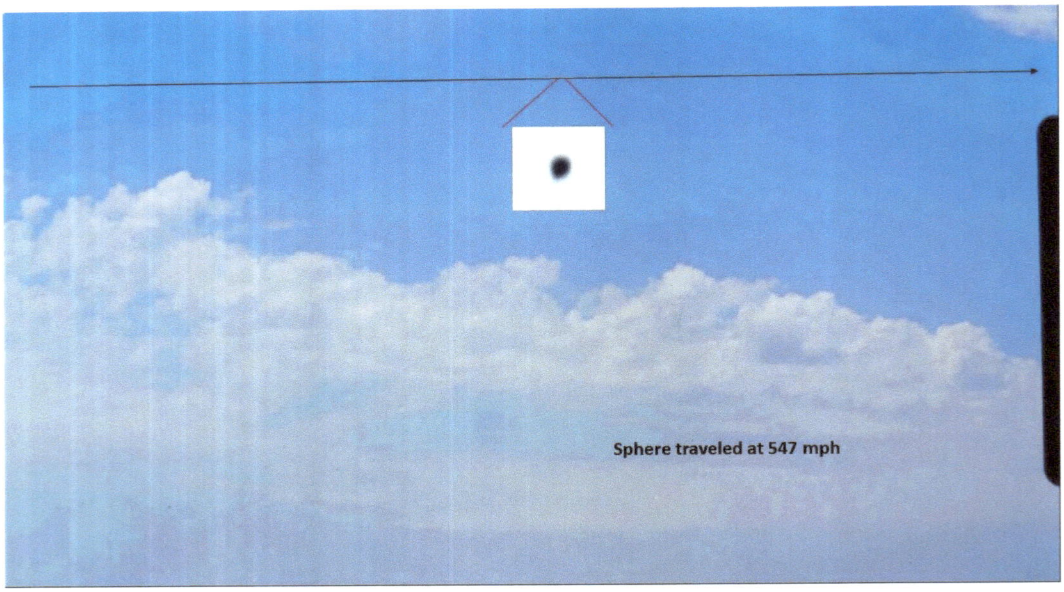

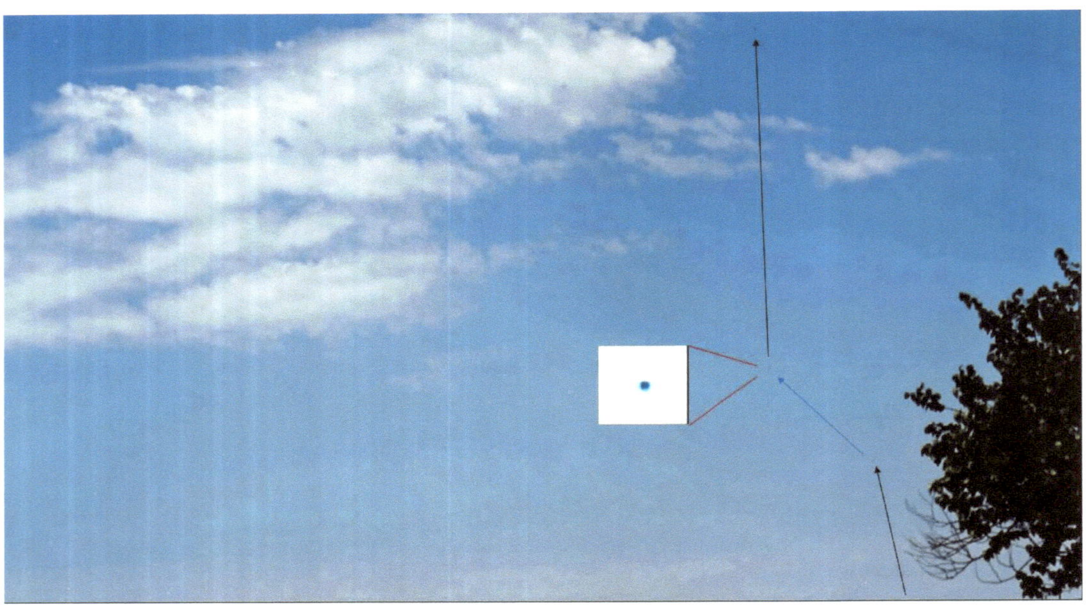

Hovering Light

Case # 110320
Date: 07/21/2022
Time: 8:33 pm
Place: Galt, MO

On Tuesday July 21, 2020, the witness, husband and a child all witnessed what appeared to be a hovering light which also moved—toward the north. After consulting Flight Radar, the object appears to be SWA663 in route to Phoenix from Midway Airport. It was flying at 18 degrees elevation, and by 8:36 pm was 23.4 miles distant, cruising at 40,000 ft. At That point it was due west of the witnesses traveling in a southwest direction which Would have appeared to be south to the witnesses. The object therefore is an IFO, man-made. Images enhanced at the right show the tapering for the cockpit, the fuselage, the shadow of the under-side of the wing, and blue color of the fuselage. Photos are on the next page…

Figure 2A typical Southwest plane

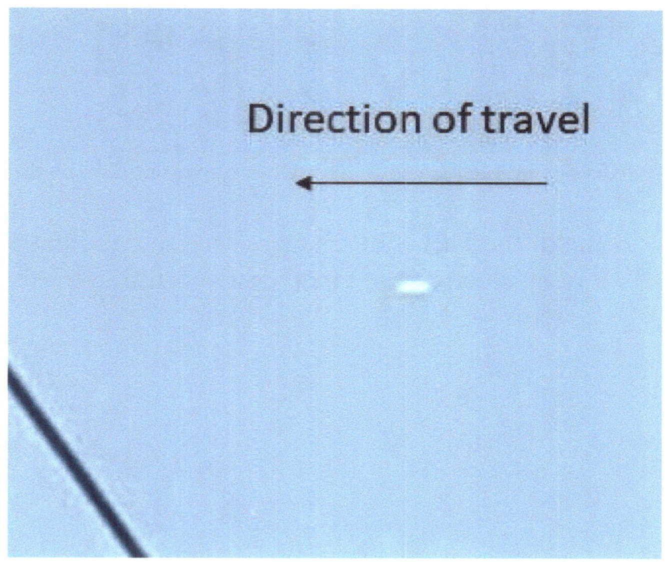

Witness' image

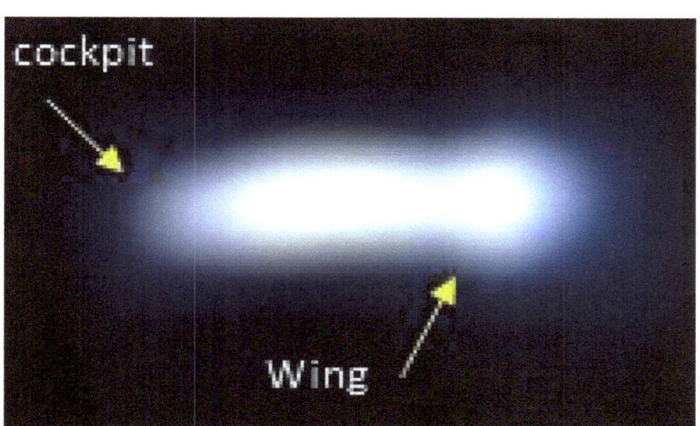

Figure 3Enhancement

Despite the object being found to be Southwest flight 663, there were more surprises in store!
There were also 7 Fast Movers.

Object moved through in .166 seconds

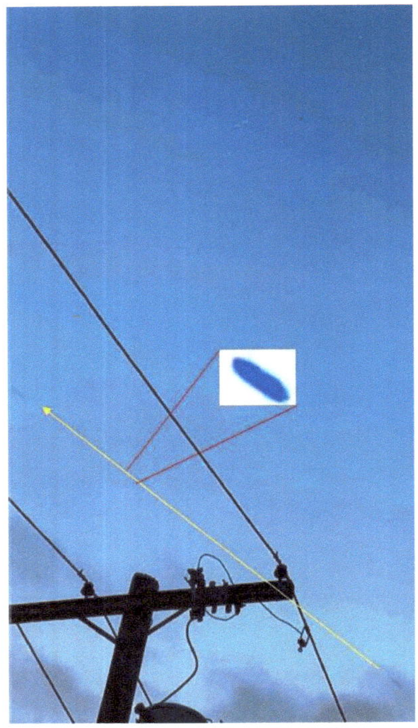

Object moved in .53 seconds.

Object moved in .3 seconds.

None of the Fast Movers were seen due to the speed by which they flash across the sky. Only by slowing the video down to frame-by-frame speed can one observe these craft.

Daytime Object

Case # 111061
Date: 08/21/2020
Time: 10:48 am
Place: Galt, MO

This is the *second* reported sighting for this witness exactly **one month** later. On Friday August 21, 2020, the witness, and her husband witnessed what appeared to be an unusual object which moved—toward the northwest from north. After consulting Flight Radar, the object appears to be a Piper PA46. It was flying at 29 degrees elevation from the witness, and by 10:59 am was 2.46 miles distant, cruising at 13,000 ft. At that point it was traveling westward and would have appeared to be northwest as it flew on. The object is an IFO, man-made. **However, once again, there were also seven Fast Movers**. Two are shown in below.

Figure 4Type of aircraft seen

Original image

Enhanced. Aircraft shape confirmed

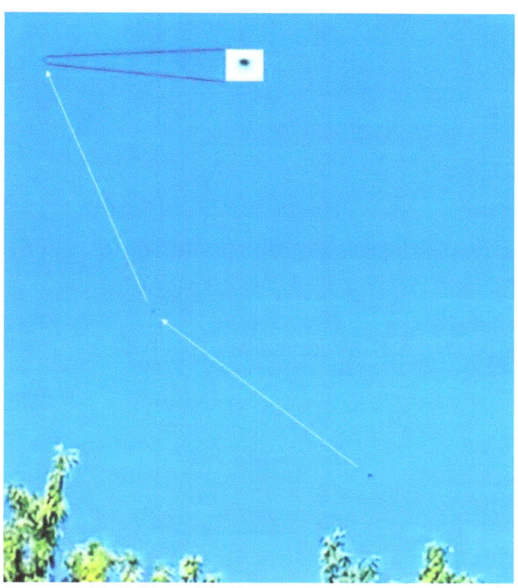

Frames 8:21, 9:02, and 9:09 traveling though the screen in **0.6 sec making a 29-degree sudden directional change.**

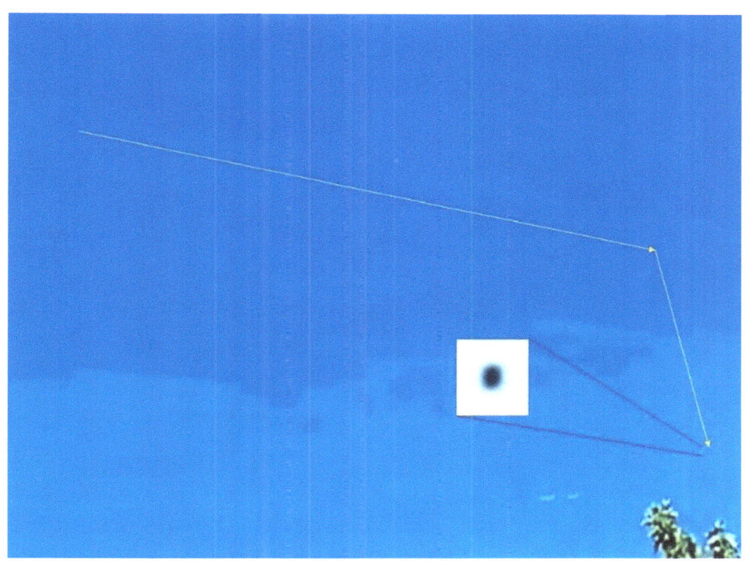

Frames 30:28 and 31:14 merged. Object traveled in .53 seconds making a **hard 63-degree instant change in direction.**

Wayne Lawrence, Lily Niva, Bill Spicer, Margie Kay

Caught During Takeoff

Case # 121200
Date: Monday 03/08/2021
Time: 3:37 pm
Place: Chicago, IL O'Hare Intl Airport

While photographing departing aircraft, a witness caught a white "tic tac" shaped object below the nose of the Alaska Airlines jet. There is an area beneath it that looks like blurred movement. The 737 jet was 102 feet long, and the area that it occupies on the photo measures about 110 feet in actual length. The plane was traveling at 150 mph at takeoff. It was approximately **1000 feet** in the air according to the witness. The distance of the jet from the witness when the photo was taken would have been **.4 miles** according to Google Earth measurement. Using the elevation and distance information above, the angle of observation of the jet at that point was calculated to be 25.5 degrees.

The speed necessary to clear out of the airspace in the photo-- at a high enough speed not to have registered in the subsequent photo--was calculated as if the object were at 1 mile distant, 2 miles distant, and 4 miles distant range between Mach 1.3 and Mach 5.2 depending on how far the object was relative to the plane.

Blurring beneath the Tic Tac object showing high speed and vertical movement

The witness took rapidly sequenced photos in bursts on either side of the photo in question. Neither frame on either side of the main image of the tic tac shows the craft. The white object seems to have been caught by the witness as it was rapidly moving in one photo only. Since only 2/10 seconds elapsed between the photo, and the next frame in the sequence at a rate of 5 fps, it seems that the tic tac shaped craft is a **Fast Mover**. The photo was taken with a Nikon D-5600 camera.

Under the Moon

Case # 124292
Date: Tuesday 08/16/2022
Time: 9:10 am
Place: Chicago, IL

A video taken by the witness was provided by an iPhone 12 Pro Max. The subject was two white sphere-like objects (not pictured). There were 6 Fast Movers also. Two are depicted here.

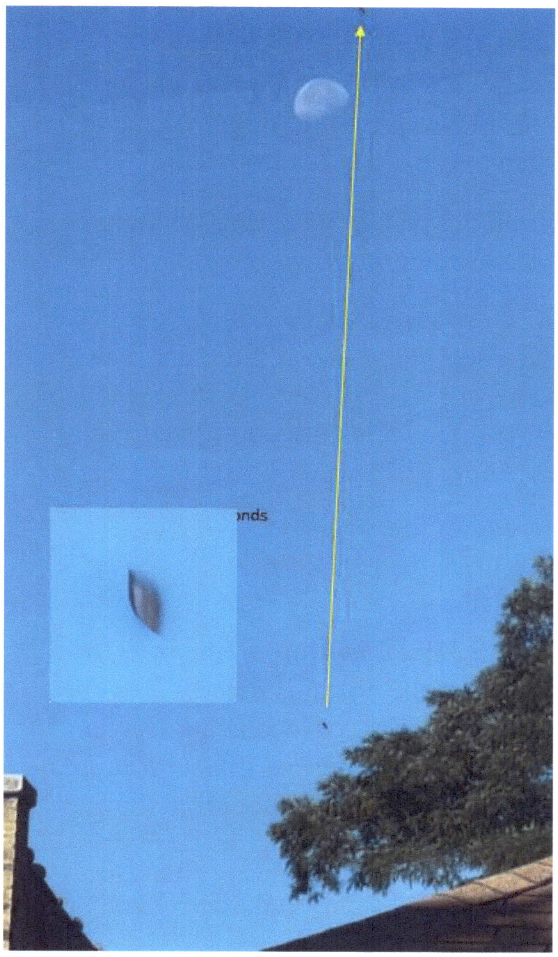

Merged frames 4:54:18 and 4:55:00
.4 seconds

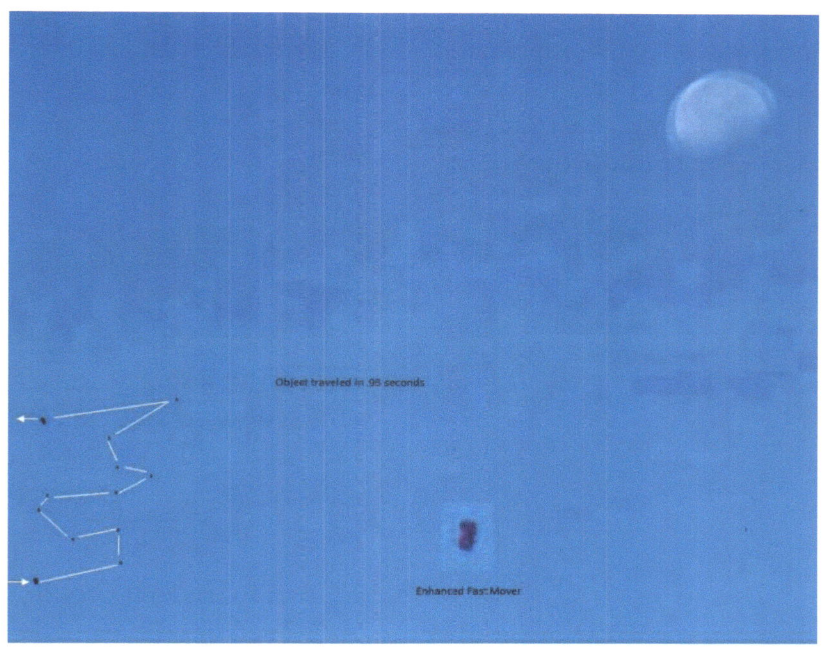

Merged Frames 4:09:29 to 4:10:27. Note zig-zag movement in .93 seconds

Skywatch

Date: Friday 06/23/2023
Time: 2121:21 CDT
Place: Naperville, IL
Reported: Wayne Lawrence

June 24 is internationally recognized as the first of two "World UFO Days". A Skywatch was organized on Friday June 23rd, 2023, in Midwest MUFON in honor of that event—which recognized the first report of UFOs by aviator Kenneth Arnold who was flying over New Mexico and witnessed a fleet of crescent-shaped craft on that date in 1947.

I waited until a little after 9 pm so that it might be easier to see lights moving. Civil twilight began at 9:06 pm, so the sky was light, but the sun was below the horizon.

First, I scanned the southern skies which presented the best view from my position. I settled on a view bordered by a wall and some trees that would give me direction and elevation information.

All told, I took three short videos with my iPhone 12 having steadied myself against a wall. When I reviewed them, it was obvious in the third video that I had captured a short UAP event. I did not see anything from the quick viewing of the other two however once I was able to download all three and look more closely frame by frame, I did see a second UAP in the first video. They are both considered Fast Movers due to their speed.

This object was seen in video IMG 2599.mov from frame 6:19 to 7:00. The flight lasted .4 secs. The shape of the object is saucer-like. The angle of observation began when it was first sighted at 26 degrees elevation and ended at 33 degrees when it vanished. It was seen in the SW at 236 degrees azimuth. The altitude and distance from the camera are not known.

It shot straight up very rapidly and then disappeared after .4 seconds. It is shaped like a type of disc.

Frame 6:29

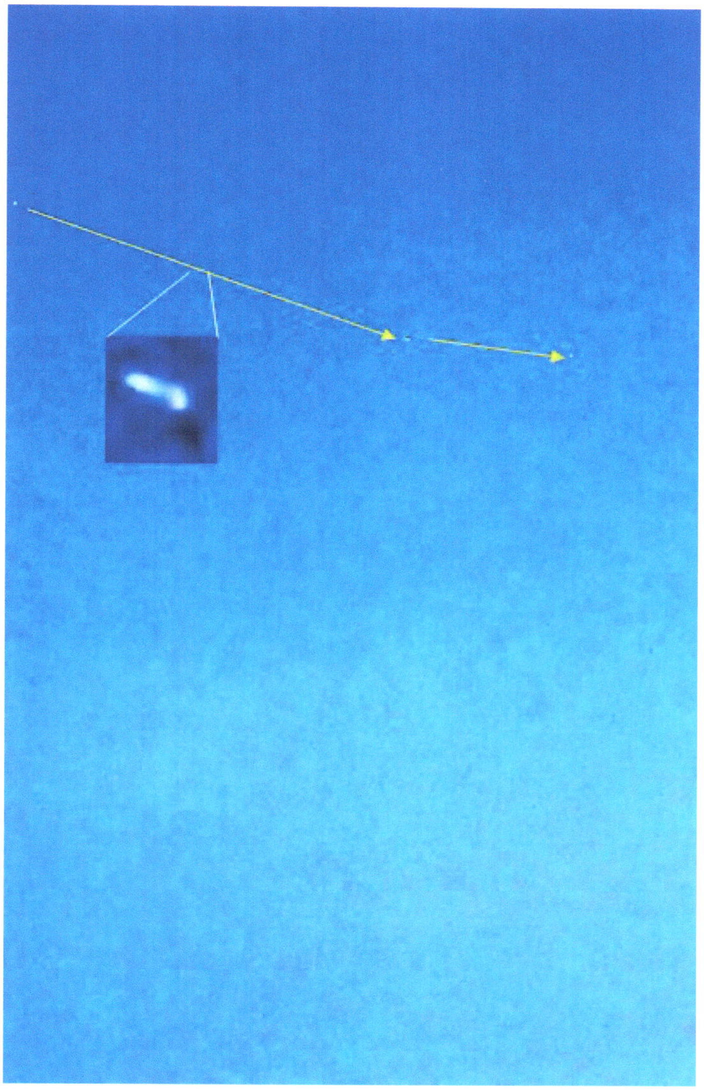

Merged frames 30:18, 31:18, and 32:00 , 1.4 seconds.

Original

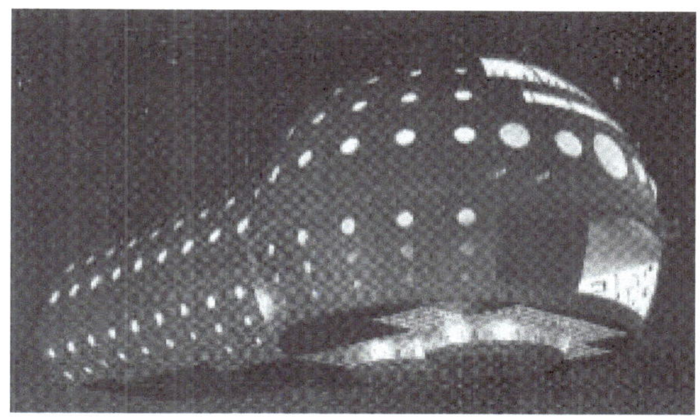

Faint resemblance

The object became visible at 35° elevation in the SSW at 205 ° azimuth. Distance from the camera, size, and altitude are not known. The craft seems rod-like with a slightly larger end at the front.

This particular UAP appears to have a cylindrical body with a larger sphere at the front end.
There is a similar artist's rendition from a witness testimony which appears in *The Alien Agenda.*

No FlightRadar 24 aircraft were at 205° azimuth, and 35° altitude. The only plane in that area was at 180° azimuth --and 5.8° altitude and could not have been seen. That was flight JIA5432.

* Image from *The Alien Agenda*, by Richard Dolan, p. 66, artist: Tom Bogan.

"There are numerous daytime and night time photographs and videotapes of clearly non-human spacecraft from all over the world; these films and videotapes have been evaluated and deemed authentic by competent experts in optical physics and related fields."

Steven M. Greer

Chapter 2
Quantum Communication

In chapter 2 I will look more specifically at the quantum communication effect that I have experienced as I look back over ten years of observations and investigations. In this chapter-- I refer to types of active communication with non-human intelligence entities (NHI). I suggest that there are several types of consciously intentional quantum communications which have resulted in acquiring contact data (photographic, or experiential). These also apply to all three chapters.

TYPES of Intentional Communication:

1. Active telepathic contact with NHI— asking NHI to show themselves and/or their craft. (Contacting NHI)

2. Recording of video to find unknown UAP/UAV. The intention is communicated generally and is sensed by NHI. (Intention communicated) 3. Non-human intelligence chooses to show themselves. Recipients witness the disclosure. (NHI showed itself)

4. Remote Viewing of UAP, locations, events, entities, or phenomena. The intention to view is broadcast. It may be sensed by NHI. (RV) 5. Random recording of video, without knowing why. Subliminal message is sensed to record a video. (Unsure why recorded)

6. Accidental discovery. Surprise to find UAP in the recording. NHI intentional activity, unintended discovery. (UAP discovered accidentally)

Quantum Communication has been described as "…focused attention…causes the percipient and the object to…merge so that…information is obtained." *

* Hernandez, etc…Beyond UFOs, p. 378.

Consciousness has been an area of developing study for several decades most recently, but which has its roots in philosophical pursuits dating back to over 2,500 years ago. Consciousness…is an essential component of the universe… The 'quantum universe' … describes the basis of consciousness: It elevates information…to the same…status as matter and energy. In other words, it links consciousness and matter, and energy. (Hernandez, et al, p. 91)

In the early 20th century, Einstein equated energy with mass and a profound change in understanding occurred. In recent decades, with echoes from earlier centuries, energy and matter have been shown to be essentially the same. Light is essentially waves of electromagnetic fields, that sometimes appear as particles. We used to think that outer space was a total vacuum but know today that (in reality) a quantum field permeates all parts of the universe. (David Tong, "Quantum Field—The real building blocks of the universe," YouTube.). Essentially-- we are made of fields. Us and the universe around us. This suggests a likely reason that quantum communication occurs is fields of consciousness embedded in fields.

Consciousness is not simply what you or I or animals experience. It appears to be part of the nature of all life, and arguably even of the physical world itself. Consciousness is ultimately the basis of everything. (Hernandez et al, p. 90)

Conscious Intention

For this notion to be accepted, as in all new major scientific breakthroughs, it is not a simple task. There are many detractors who are steeped in older concepts of science who want to maintain a status quo. History has amply demonstrated that science progresses by funeral, not by reason or logic alone. *The Conscious Universe,* Dean Radin, PhD, p. xxi.

The unique nature of **remote viewing** involves conscious intention as a form of quantum communication. One or more of the above six types of communication apply to all UFOs/UAPs in this book in which the **intentionality** of quantum communication has been involved-- and some anecdotes will be shared. The information is presented to show new aspects of our universe which we as a species need to come to grips with. Hopefully, we can begin to internalize the new understanding of the many NHI around us that are very real. Hopefully, we may begin to make the mental, social and behavioral adaptations that will reflect our "growth" as beings.

We are living in a galactic neighborhood that is much more crowded than we have supposed. We also live in a planetary neighborhood that is much more crowded than we have known. It is high time that we met our (*mostly invisible*) neighbors and begin the process to acknowledge this new reality! This is as important a step as was the understanding that came from Darwin's

On the Origin of Species, or even the discovery that the world was not flat. The great question is: do they want to meet us?

In my own experience, I have occasionally taken videos to **intentionally** see if there were any **fast movers** present. Thus far, they have been present 100% of the time. I theorize that I have broadcast the intention of the videography-- alerting those non-human intelligences (NHI) who are nearby to appear if they so desire. As I look back, I realize also that many times I decided to capture video or snapshots very intentionally because I felt prompted to do so (*without realizing why*).

My study of the quantum effect in communication has led to a new understanding that is emerging in my own experience about the many new layers of existence that I had not previously been aware of. Interestingly, I am not at all surprised by them. In fact, it all seems to be very natural.

Quantum Communication Conscious Intention

A Brilliant Sphere

Case # 59513
Date: Sun. Aug. 31, 2014
Time: 8:15 pm
Place: Lee's Summit, MO

Type 3
(**NHI** showed itself)

On Sunday night August 31, 2014, I was watching television in our great room and out of the big windows that open to woods behind our house. I saw a light so unusual that I jumped out of my chair to look at it. I realized that I had seen it about two nights earlier just very briefly as it passed by… but at the time gave it no thought other than it was characteristically unusual but did not stop. That time I jumped up too, it was so odd.

The spherical light appeared at 8:15 pm by clearing the trees. In fact, I could see that something was preparing to emerge, it was so bright.

Photo of sphere

Adjustment to see tree line

The light hovered for about two minutes and was constantly lit. It came from precisely 340 degrees, or NNW, and just floated in place. It likely emanated from a cove 1.07 miles away... It did not seem to be exceedingly far away. The angle of observation was 15 degrees from the horizon. Almost anywhere between us and the object would place it over the undeveloped western shore of Blue Springs Lake. It was a brilliant white light unlike anything I have ever seen, but then moved away to our left (to the West) in the general direction that would lead to Lakeland Lake, and out of sight. When it turned **westward**, it appeared to split into two lights. The front light was very bright yellowish white, the second was red. The lights stayed on, but did seem to pulse slightly. This westward corridor has also been used by other UAPs observed since 2014. It is undeveloped of any housing and seems to be the direction of choice for other UAPs when they travel.

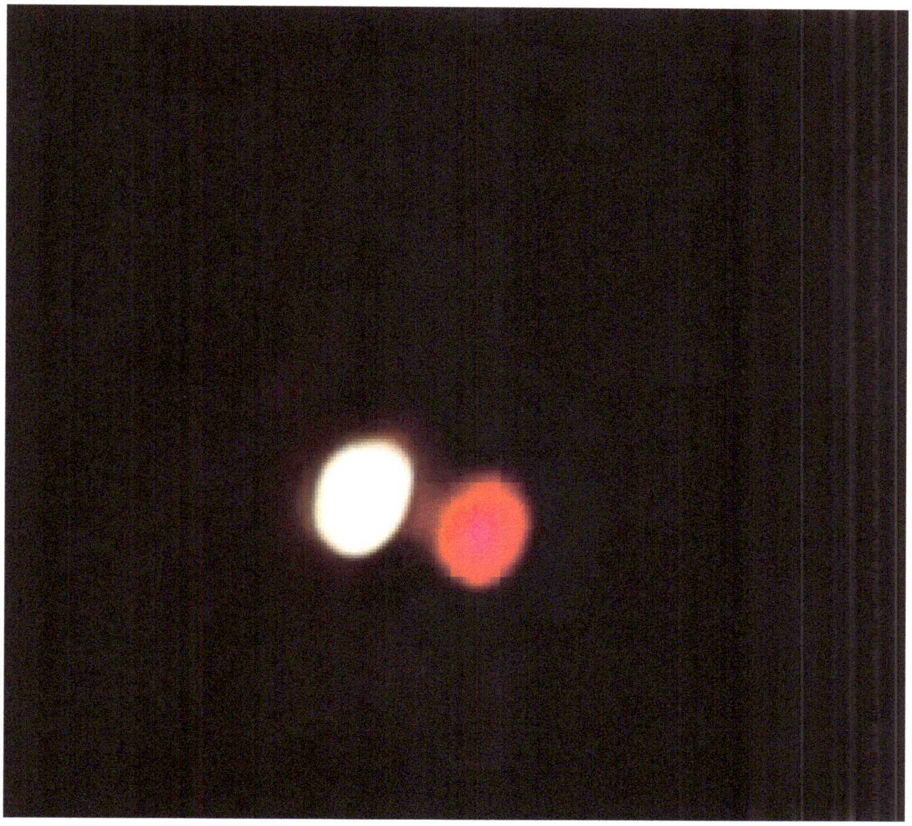

After the sphere morphed into two.

IMPACT:
I was not looking for a craft. This sphere was directly in front of me. This was my second sighting of a UFO/UAP and it kicked off my interest. The reason that I believe that this sphere was shown to me is that it stayed for about two minutes just hovering…almost as if it were wanting to be sure that it was noticed. In fact, it was a perfect setup to be seen. It could have appeared anywhere else in 360 degrees. Instead, it was directly in front of me. We were the last house built in that part of the neighborhood and there was a lot of undeveloped land around us that would be built up in subsequent years. Could it have been just dumb luck that I saw it? Perhaps. At the time, I had the distinct impression that I was supposed to see it.

See Chapter 3, last RV session, for more detail.

Quantum Communication

Conscious Intention

Four Cases of Meteor-like Objects in the Same Area

TYPE 3

NHI showed itself

1. **Meteor-like Light that Hived Off Into 4 objects #59813:**

On Friday night **September 12, 2014,** at 11:52 p.m. a bright object appeared in the northwestern sky. It appeared at first to be a meteor, except that it travelled more slowly and lingered longer in its descent than normal meteors do in just a fraction of a second. The bright light appeared next to "hive off" in a straight line until 4 objects were now continuing the downward path and then disappeared below the tree line. The total time that elapsed was about 3 seconds.

The light was in the NNW part of the sky and travelled downward to the right. It looked to be approximately 1 mile or so distant. I judge the size of the objects to be somewhere between 10 and 25 feet, although that is exceedingly difficult to know for certain. Each of the objects was as bright as the original and the downward flight path did not deviate from the slope it originally had begun, roughly 45°.

What is especially interesting, is that these lights descended to the within 500 ft of the same exact place that the bright sphere emerged from on August 31 at **340° azimuth**. (See preceding case # 59513). The odds of that happening are **332,212 to 1**.

Most meteors burn and are visible very briefly as the air molecules ionize. They occur high in the meteoric area of the atmosphere because they burn up between approximately 75 miles and 45 miles altitude. This was the **second** object in the same area since May 1.

2. Meteorite-like object # 60680

TYPE 3

NHI Showed itself

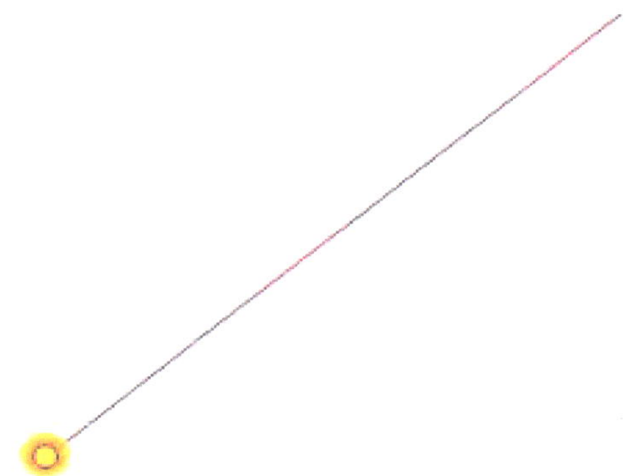

Thursday May 1, 2014, at 10:45 pm, I had been watching some blinking triangular shaped lights with binoculars. I went looking for my camera and tripod.

After trying to see if I could take a photo, I gave up as the lights seemed to have gone. I continued to scan the sky with binoculars and saw what looked like a meteorite burn through the atmosphere not far from the lights. It entered Blue Springs Lake at about 13.5° azimuth, or North, and the angle of entry was about 35 to 40 degrees from the horizon . The area in the sky from which it was first noticed was at about 60° azimuth.

The descent of the object lasted longer than normal—perhaps 2 seconds or so and had more fire in the trail than I have ever noticed in meteors before. There was no sound.

Including the Brilliant Sphere on Aug. 31, and the Meteor-Like Light that hived off into four objects, this makes the **third** object to have entered **the same general central area of the same Blue Springs Lake.**

The odds of three celestial objects falling in the same area within feet of each other is becoming astronomical.

3. Fiery Trail of Light #60680:

TYPE 3
NHI showed itself

In the early hours of Wednesday, October 15, 2014, at 12:06 a.m. I was looking at various lights that had been moving but not able to capture on video. I was using my binoculars, and placed them down for a moment, only to notice the last full second or so of a fiery bright yellow/orange trail of light that dove at about an 80-degree steep slope directly into the Blue Springs Lake.

It would have entered at about direct north, or 360 degrees direction. Depending on its distance away, there would have been many places in the lake that it could have entered. My estimate places it at about a mile from my observation point at home. The magnitude of brightness was at least -10. This means that it was as bright as a half-moon. The frequency of such sightings is **once in 2 years**.

This now makes the **third** object of a "meteor"-like nature in 6 months and the fourth overall . The trajectories of all three that I have seen since May have gone to the same place within yards of each other. The statistical probability of this occurring in the same spot on earth is so high that it nearly defies calculation. The calculation however places the odds to be about **36,663.85 x 10^{12} to 1**.

Within 2-3 minutes, two planes headed directly toward the same spot. One overflew the Blue Springs Lake and the other on a parallel and slightly tangential course just to the western side of the Lake. Neither plane was seen to return after flying over.

4. Brilliant Green Streak #64349:

TYPE 3
NHI showed itself

Shortly after midnight of March 30th, 2015, (early Monday morning) at 12:39 a.m., I noticed a brilliant green object descend in about a second of time from an angle of about 80° in the night sky. It appeared to be like a meteor; however, we know that this is not possible under 12 miles altitude. This was seen through our living room window.

The direction was 333° or roughly NNW. The object appeared at about 17° altitude from the horizon, and when it had descended to 12° elevation, it hived off at least one smaller object that continued in the same direction as the main larger craft. It just appeared.

The objects would likely have entered Blue Springs Lake at about **343° azimuth, nearly North.** At approximately 1 mile distance, and covering a descent of 834 feet in about a second before vanishing beyond the trees, the objects would have been travelling at a rate of approximately 569 mph.

This represents the **4th** such objects in nearly the same area as the three previous meteor-like craft from May 1, September 12, and October 15, 2014--and now March 30, 2015. (See photo of area on next page…)

Five objects overall have been seen coming down into the lake in the same area, including the Brilliant Sphere from Aug. 31.

One can only wonder about what attracts them to the same area. Two recent books about flying objects in Earth's oceans and lakes have catalogued many of these underwater incursions. They are: *Unidentified Submerged Objects*, and *The Alien Colonization of Earth's Waterways*. Very little is known about them, but many possibilities have been raised. Some range from using hydrogen from water, to harvesting foods from the seas, to establishing bases underwater and many more.

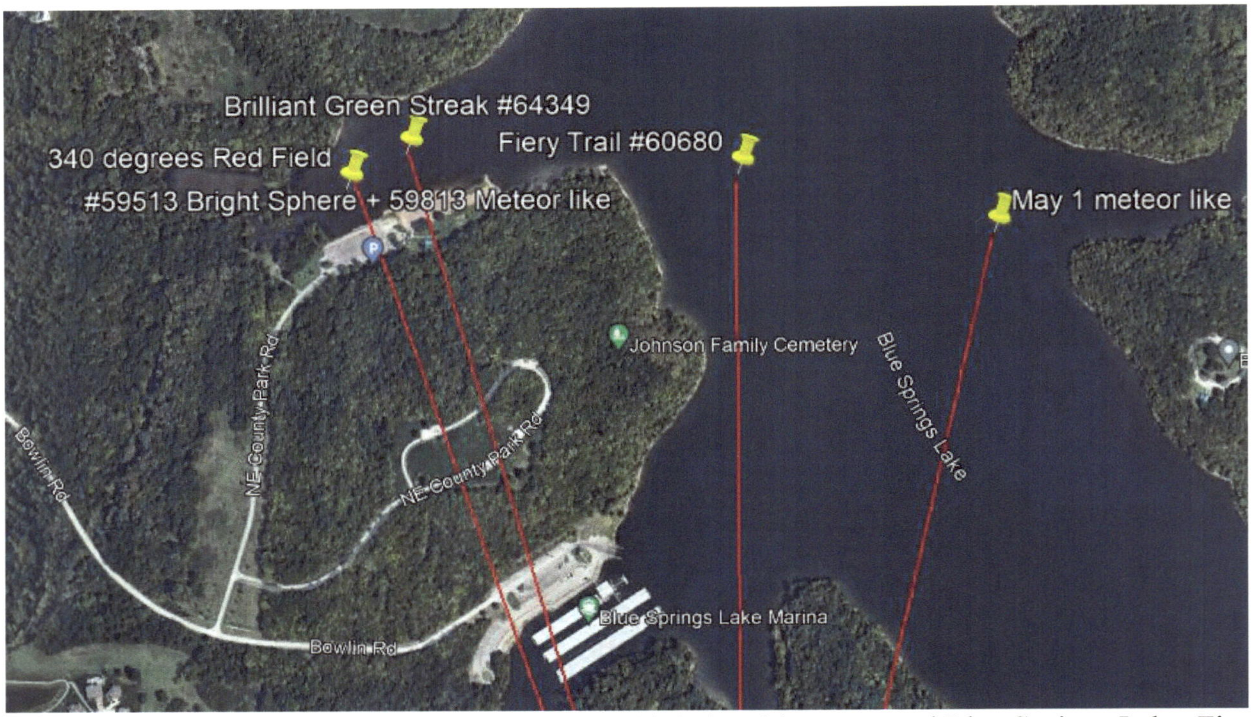

The above Google Earth Map shows the locations in which the objects entered Blue Springs Lake. Five of the objects 5 objects have been noticed entering. F**our** were within feet of each other—but they all spread through the middle of the lake.

SUMMARY:

What are the odds that the same observer would be at the same place, at the right time, on the right dates…

- …that **1 unusual bright sphere** ascended and hovered…
- …that **4 meteor-like objects** descended…
- …all within feet of the same location? (***4 out of 5*** *in fact were within the same area at 340° azimuth.*)

The odds are staggering for the 4 meteor-like objects to have been all at the same location-- as the odds against that calculate to **404,639.1 x 10^{22} to 1**. Coincidence seems unlikely that all **5 objects** occurred at the same area, or that they all just "happened" to be observed each time by the same person.
The craft in these cases suddenly appeared and were briefly seen. It could be that in the deceleration process, the craft become visible. If NHI pilots were involved, they would know this. Given my previous experience with UAPs "*showing off*" to me, this feels similar. Three such cases are explored in the next few reports about a "Flying Triangle," 2 incidents with "Starburst-Shaped Craft", and "Telepathy". It could also be that the pilots of these craft were just going about their daily lives. Whatever the case--there is just a lot of synchronicity going on!!

(*Non-Human Intelligence show themselves/their craft.*)

Quantum Communication

TYPE 3
NHI showed itself

Flying Triangle Near Lake Jacomo

Case # 62489
Date: Mon. Jan. 5, 2015
Time: 6:12 pm
Place: Todd George Rd area

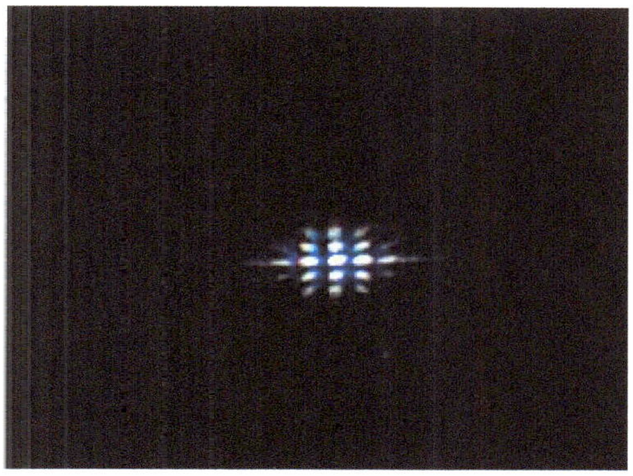

Starburst Light that left my neighborhood

On the evening of January 5, 2015, at about 6:12 pm, I noticed a very bright light "starburst" light with many rays that were pulsing as it stayed bright and moved out of our neighborhood in a southwesterly direction. It continued to travel in that direction while I was on Woods Chapel Road heading west toward Highway 291. It seemed to have continued in the distance and then I lost sight briefly until I was nearing the intersection of Todd George Rd at the stoplight.

Apparently the lights doubled back and drew near to my location at the stoplight, and they seemed to just float at about a 25° angle from the horizon, somewhere between a ¼ mile and a ½ mile away to the south.

They had changed from starburst light to two bright round horizontally oriented lights. And they came toward me then stopped about 2000 ft away and hovered.

I had to stop watching them as the stoplight turned green, and I proceeded to cross the bridge so that I could get onto Highway 291 and travel south toward the heart of Lee's Summit.

When I was in my lane on 291 south, I noticed the two lights had followed me and were also travelling in a southerly direction. But now the two lights were vertical to each other. They were **½ a mile** away. They were at an angle of about **35°** from the horizon.

Two airborne lights

The object then changed its position in the sky and the two lights became three --as I realized I had seen the edge of a triangle-- because the third light had raised up from behind the other two lights. Red, blue, and green lines of light joined the corners of the triangle. **Each time that I looked back to the triangle, it changed its position and moved around**.

Finally, as I reached an area that was inhabited again, the triangle disappeared. This was the first time that a craft kept changing its position in the sky each time that I looked at it. It was also the first time one followed me. Its occupant(s) wanted me to see it.

The occupant driving the object then changed the position of the two lights to a vertical orientation. It moved in various positions, as if it was dancing.

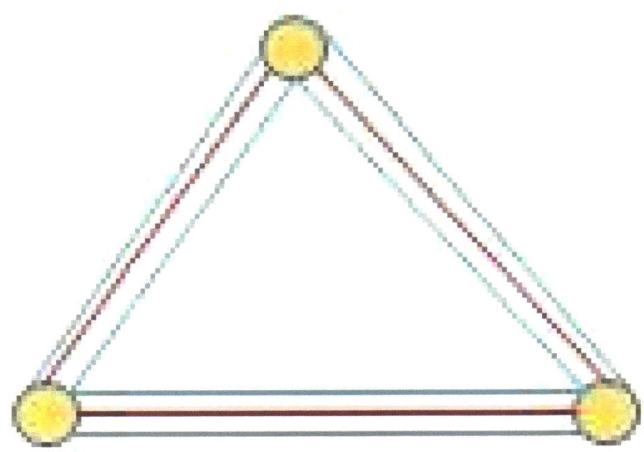

Final maneuver of the object

Quantum Communication

Conscious Intention

Starburst -Shaped Craft*

Case # 85696
Date: Fri. Aug. 4, 2017
Time: 9:30 pm
Place: Lee's Summit, MO

TYPE 3
NHI showed itself

On Friday August 4, 2017, at 9:30 pm, as I was returning home from a meeting, I saw a very interesting craft. It occurred just as I was entering our neighborhood. I thought that my dash cam was recording, but evidently it was off. I observed it travel in a straight line towards the north until it became too faint, roughly about 35 seconds.

The craft travelled in a northerly direction with both end starburst lights to the left and right of a red sphere. I estimate the altitude to have been around 600 ft. The initial angle of elevation from my perspective was approximately 25°, but then as it receded 1.5 miles to the north, it began to drop slowly until I could not see it.

I had the distinct sensation that it was showing itself.

**This sighting was experienced by the author.*

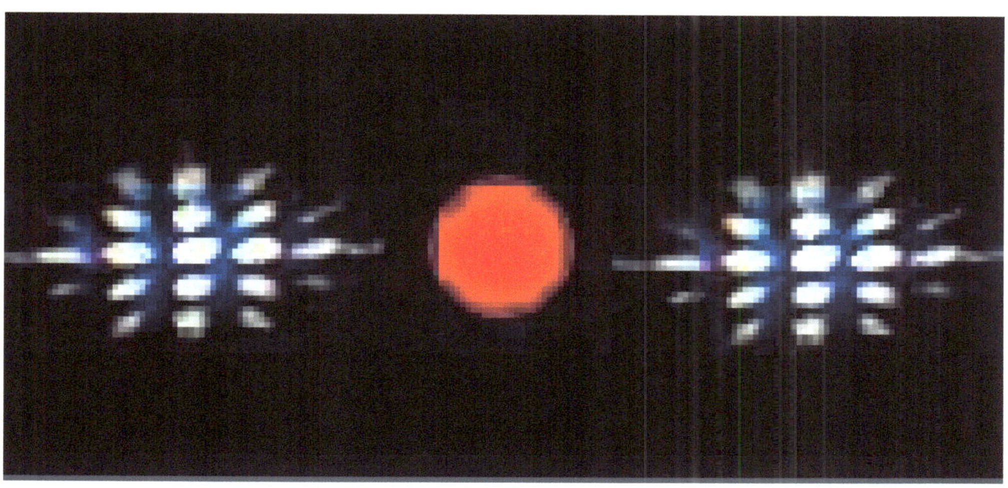

Created from an image taken Jan. 27, 2015, of several starburst lights.

One week later…

On Friday night August 11, 2017, at 21:30 DST the craft returned to my home.

My wife had closed the blinds, and I could not see out. At 9:30 pm, I recalled that it was precisely one week earlier that I had seen the odd craft with two starburst lights on the ends and the red light in the center. I intentionally had the following thought: "***If there were another fly-by tonight by the same object, I would not be able to see it because the lower blinds are closed.***"

Perhaps 5 seconds later after that thought, the object must have intended to be seen because I saw the craft in the upper picture windows of the living room as it flew by. This time, it was in a NEW configuration…with starburst lights on the bottom, and a red light at the top, in a triangular configuration.

It felt like the NHI had read my thoughts and wanted to be acknowledged as a different shape. Instead of flying across the lower windows like it had done one week earlier *(which were closed)*, it flew across the higher set of windows! It chose to show itself again.

SUMMARY:

Twice in one week's time, the starburst craft had shown itself at the same time of night on the same day of the week. The odds of that happening at any given minute within a 24-hour period times 8 days is at least: **184.88425 x 10^{23}** to 1.

The mere fact that on August 11th (eight days later), the object reacted to my telepathic thoughts within 5 seconds to show the starburst craft in the **upper** windows-- is itself amazing.

Quantum Communication
Conscious Intention

Disc-shaped Craft

Case # 77979
Date: July 21, 2016
Time: 3:28 pm
Place: Lee's Summit, MO

Type 5
Unsure why I recorded it

On Thursday July 21, 2016, I took a random video that lasted 39 seconds at 15:28 DST. During the video a flat or possibly disc-shaped craft travelled from frame 18:35 to 19:05, or ½ of a second. The angle of rapid descent was at 14° from the right side of the frame to the left. Nothing else was seen in the remainder of the video. The angle of elevation when it entered the screen was 12°, although it clearly came from much higher.

The direction of initial view was 350° or nearly North, and it disappeared from the camera view at 336° or NNW-- although it continued in a straight line ending to about 320° or NW where it likely ended its flight. It disappeared on the west side of Blue Springs Lake, interestingly, into another much smaller lake.

ANALYSIS:
In 16 frames, the craft covered a distance of .3 miles in ½ of a second, which converts to **2,160** mph. Most likely it entered the small body of water, although not actually seen doing so. The direction of the 14° descent, by projection, ended NW at 320°. This small lake is about 1600 feet west of Blue Springs Lake. The size of the object appears to be close to 35.4 feet.

The video evidence shows a craft flying at an incredible speed of 2,160 mph-- fractions of a second before ending its travel within 3/10 of a mile. This demonstrates a control unknown to our engineering. The physics of such flying makes it impossible to do with known technology if it were a human vehicle.

Frame 18:47

Figure 5Enhanced view of frame 18:47

SUMMARY:

This was a spur-of-the-moment video--yet I am unsure why I felt that I had to record it at that time and day. Once I started, I was very intentional in wanting to set up the camera on a tripod, by carefully selecting the direction, and the setting of the zoom lens. I could have chosen many directions and elevations. But the particular choice that I made allowed the flight of the disc to be captured on video.

As with many other occasions of intentional and deliberate preparation, there was a direct relationship with a very unusual sighting and capturing of images. The odds **against** just happening to record this craft by luck are exceedingly high.

Quantum Communication
Conscious Intention

Telepathy with a Craft of Red and White Spheres

TYPE 3
NHI showed itself

At about 5:55 pm on Monday Nov. 8, 2021, I was driving west on I-470 highway a little past Douglas St. when I saw an **oddly shaped** craft fly in front of me at about 20 degrees elevation. It crossed over the highway from right to left.

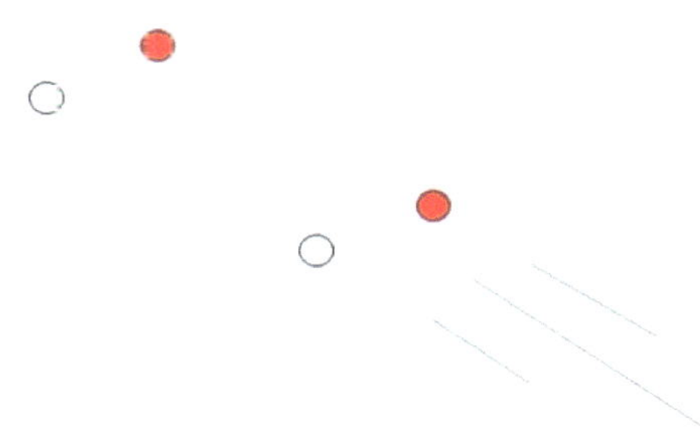

View as craft crossed in front of me

Located about ¾ of a mile on the west side of Douglas, it crossed Highway 470 about 50 yards in front of me. It came from the direction of Unity Lake No. 1. Outside, it was civil twilight at that time. The sky was still light all over, but the sun was below the horizon.

What was especially odd about it was that on the right side of the craft there were two red lights and two white lights on the left side. There was a space between them from front-to-back, but they all seemed tied together as one craft.

There was no superstructure between the lights. At one point it banked to the left and then straightened out. I had to keep my eyes on the road.

But the next time I looked, it had changed position (as if it knew I was looking at it). Now the white lights were on the top, and the red were below them. Then I lost sight of it as I had to keep driving. It seemed to have resumed whatever journey it was on.

As I drove, I wondered where it was going, and if I had totally missed seeing it ever again. I got a mental image of it arriving somewhere in the direction that it was seen traveling—but it was hovering.

A crazy thought entered my mind…that perhaps I could see it again somewhere along my destination.

I turned from 470 onto View High Drive and started south toward my apartment. I knew that there might be an opportunity to see it again unobstructed as I neared the top of a hill in the Longwood area.

So, I sent a message telepathically in hopes to see it again.

Sure enough, it was hanging around apparently just waiting in one spot. As I mentally sent the route that I would take to the apartment, it got closer and when I hit the bottom of the hill, it was parked above a tree 500 yards away. This time I did not see it in a sideways profile but **head-on**.

A strange thing happened next. I should have taken a video or a photo but strangely didn't. I didn't linger either and the thought occurred that I might see it better overhead in the parking lot.

So, I moved the car, but then when I got there, I realized that the view was too obscured by the streetlights, so I turned back to see if it was still above the tree-- but it had left. Then I saw it going over the parking lot to the northwest and into the distance.

(Note: It was very *odd that I didn't even think to take a photo!!)*

Bottom of the hill in the apartment area

SUMMARY:

The mathematical probability of the oddly-shaped craft crossing the highway just in front of my vehicle, at the precise day and time that I was there and then responding to my telepathic message by turning from its path and meeting me at the bottom of the hill at the apartment complex—defy calculation.

This is definitely a case of quantum communication with conscious intention.

Quantum Communication
Conscious Intention
Synopsis

OVERVIEW:

1 brilliant sphere, 4 meteor-like objects, a triangular craft, a starburst craft (twice), a disc-shaped craft, and a craft of four lights all presented themselves using some form of quantum communication with conscious intention.

- ❑ Three of the cases involved telepathic contact with the NHI.
- ❑ Six cases were direct results of intentionally showing themselves.
- ❑ One case involved being unsure why I was prompted to record a video and discovering later that there was a UAP in the recording.

SYNOPSIS:

These ten cases involved conscious intention by NHI to pilot their craft, in some instances showing themselves, in others perhaps unwittingly doing so.

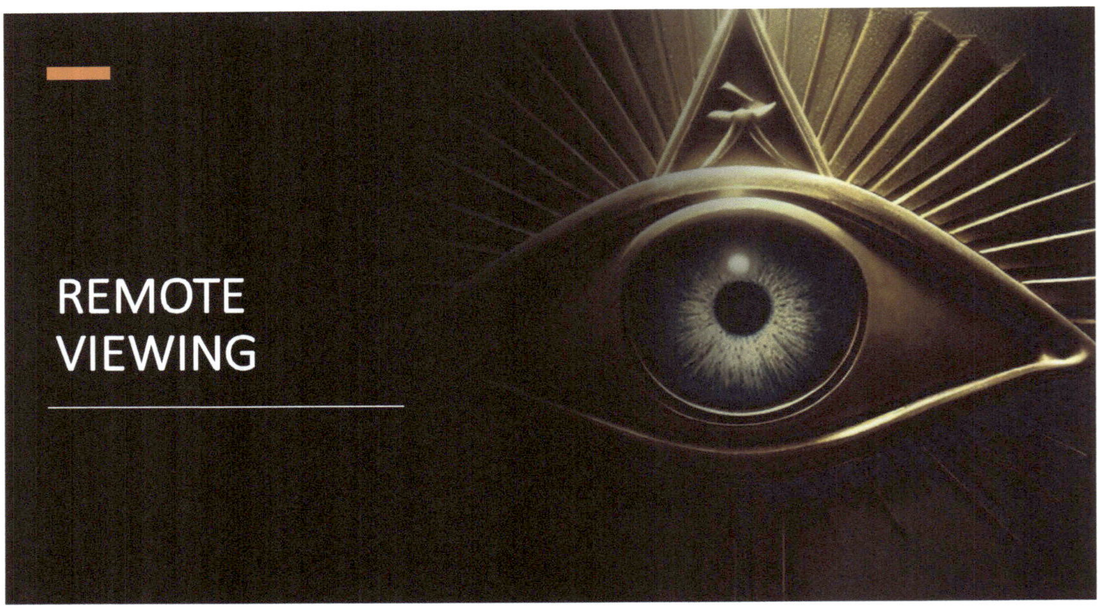
REMOTE
VIEWING

Chapter 3
Remote Viewing

<u>This chapter is about…</u>

My participation with three other Remote Viewers as we documented our individual and collective experience in viewing 11 different cases. This project has demonstrated that 70% of the cases that I have participated in had elements that were correctly perceived and verified either by photos or reports by other participants in the remote viewing exercise.

<u>What is Remote Viewing</u>?

Extrasensory perception (ESP) is synonymous with clairvoyance, remote viewing and remote perception. Simply stated, it is a way of visiting a location, or individual, or phenomenon in the past, present or the future.

One does not travel physically to the location, but one accesses information. Remote viewing is <u>defined</u> as **"information received from a distance, beyond the reach of ordinary senses"**, and occurs while the viewer is awake and focusing attention on the data target. Images of a location, objects, or of events can be "seen" by the remote viewer.

Personal evidence of humans perceiving information from a distance dates to prehistoric times. "…both experiencers and philosophers, from Plato and Aristotle forward, accepted that both physical and non-physical realms of reality must exist."

Remote viewing is considered to be one of the "psi" abilities that allows the viewer to obtain information at a distance irrespective of time factors. It is an intentional action to view a location, time, objects, or series of events. It has been used by various militaries around the world, as well as to assist in crime solving, and for many other reasons.

As far back as July 1974, the Central Intelligence Agency began to work with a Remote Viewer through a research facility in Menlo Park, CA. "*This was a remarkably accurate case, one of hundreds…*"*

"PSI" is defined as a neutral term to represent all ESP type and psychokinetic phenomena.
"*Is there any reason to believe that "clairvoyance (...aka remote viewing) is a genuine ability that most people can use if they put their mind to it? After 110 years of experimental research, the answer is clear: Yes, it appears that various forms of extended perception are authentic and are probably distributed among the general population…*"
The Conscious Universe, Radin, p. 16.

The following cases of RV communication are real.

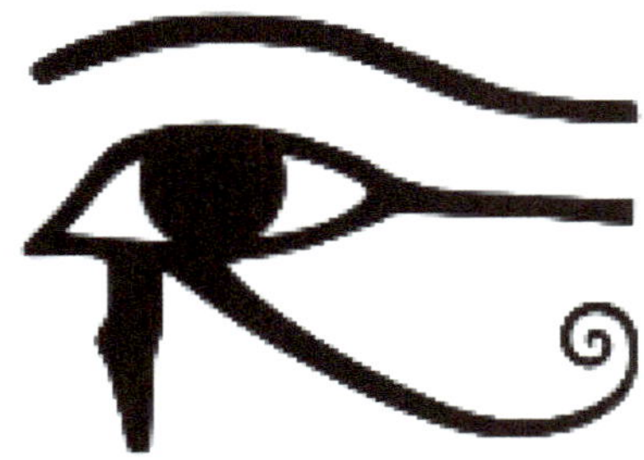

NOTE: The Eye of Horus is used in these pages to recall the function given by the ancient Egyptians to it as a symbol. It has symbolized protection, and prosperity for millennia. It has also been used metaphorically as the "Eye of the Mind," "Third Eye," "Eye of the Truth or Insight," or the "Eye of God Inside the Human Mind." It is regarding the "Third Eye" that we trace significance here. The Third Eye provides perception beyond ordinary sight. It often symbolizes a state of enlightenment. The third eye is often associated with visions, clairvoyance, precognition, and out-of-body experiences.

For this reason, the Eye of Horus has been chosen as a thematic image of interest.

Remote Viewing

A Startled Entity

RV Session 1
Date: 02/25/2021
Case # 113720

On February 25, 2021, four Remove Viewers were given the task to remote view case 113720 without any other information or access to any data. At the conclusion, we compared our notes. We were given two opportunities to view the case.

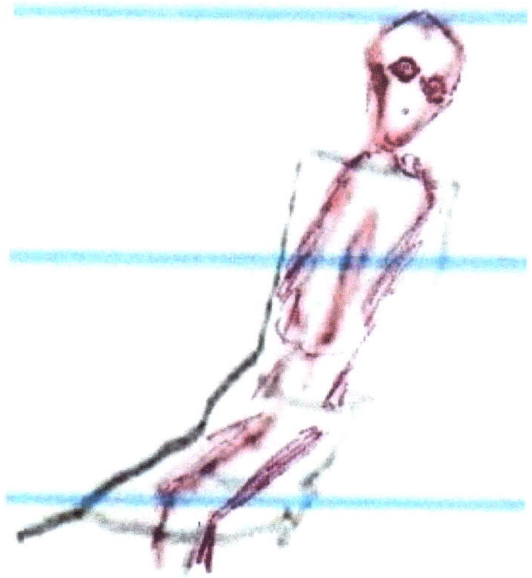

Quick sketch at end of session

1st RV Session:
I saw a NHI (non-human intelligence) with leathery skin around the eyes, which did not look humanoid. The face was slender but wider at the top. It seemed to notice me with a start, and it seemed like it was angry in the eyes. Then it got harder to see much else. I thought out: "*We mean you no harm, we're just curious about you*". At that point it relented.

2nd RV Session:
Very quickly I saw a full body view of the entity sitting in a lounge type of reclining chair, though it was not like any of ours. It was minimal. I noticed a reddish and at times an orange glow in the room. I also saw many eyes, different kinds and shapes. I also saw a shadowy figure with little detail.

Sketch with more time to perfect

Commonalities with other remote viewers:

A	B	C	**My detail:**
Other hybrid children. Girl taken upstairs.	Many other small beings	Girl unafraid.	**Many eyes, various kinds and shapes** (belonging to others aboard the craft)
		Tired-looking man. Bags under eyes.	**Entity with leathery skin around the eyes. Did not look human.**
		Shadowy person next to man.	**Shadowy figure nearby with little detail.**

Note: I had telepathic contact with, and observation, of a NHI in its environment on its own ship. It exhibited great surprise, even some anger at the intrusion, but it relented when I explained my curiosity about their ship.

Points of commonality with the case data:

- All four remote viewers saw evidence of children on board the craft.
- All four remote viewers saw that there was a craft of some type.
- Viewer "A" noted the time at 6:50 am.
- Viewer "A" said it was **Cylinder** shaped, sun reflecting.
- Viewer "B" noted a **circular** object.
- Witness described the time at 6:50 am.
- Witness described the craft as a **sphere**, and very bright.
- Two viewers noted an entity (NHI) with "tired" or "leathery" eyes.
- One viewer telepathically contacted the entity (NHI) inside its ship.

There were six points corroborated by remote viewers and/or the witness :

- ❖ All remote viewers saw children on board.
- ❖ All remote viewers saw a craft.
- ❖ Two remote viewers noted an entity with tired or leathery eyes.
- ❖ One remote viewer corroborated the witness' time as 6:50 am.
- ❖ One remote viewer established telepathic contact with the entity above.
- ❖ Two remote viewers saw a shadow figure next to the NHI with tired eyes.

Remote Viewing

Pole of Energy

RV Session 2
Date: 03/04/2021
Case # 109347

On March 4, 2021, four remote viewers were given a case #…109347 to view with no other information. The following impressions were recorded.

My Initial Observations:

- I saw a circular area that reminded me of a crater in that it was round. It could have been an object too. There was an object in the center that at first I thought was a rounded hill. But then it **looked like a pole, or energy**. (*See drawing*)
- There was **green** surrounding the area.
- I also saw a round **spherical** or **disc-shaped object** that was on the ground, with an opening doorway, and a ramp going to the ground. (*See drawing*)
- There was a rectangular object with soft corners. (*See witness photo*). What I saw looked to be at the **same** angle as in the photo from the witness.

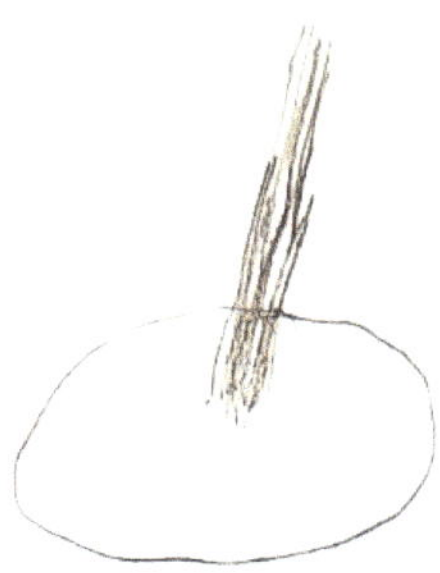

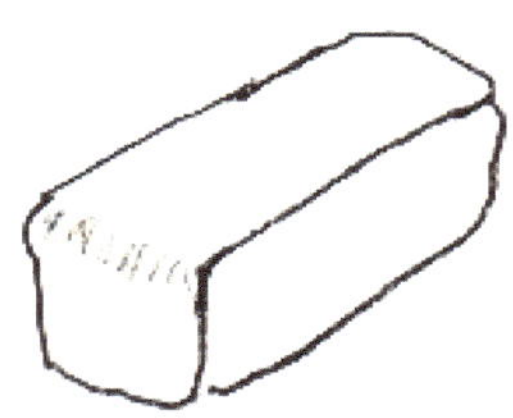

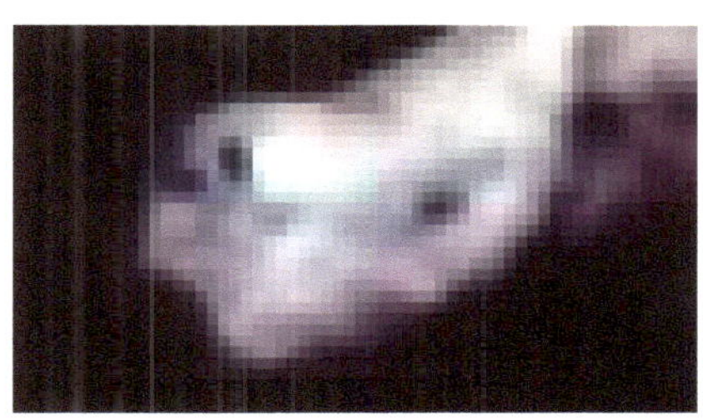

--Pole with energy
--Ship with band around it and ramp

- -Building w/rounded corners

Photo

Impressions from Remote Viewer **"A"**:

- *"There is a large, rounded **tubular vertical structure** in a Gold/white color coming out of the ground. The structure is about 3' across and 5' tall.*
- *There is a **lot of grass around** - tall grass nearby which is brown and **green** in color. "*
- "At night I see a thin white laser-like **beam being projected** from the top of the object straight up into the sky and there is no one around when this occurs."
- "A **building to the right**, small, white or tan, with a roof on it."

RV "A" drawing of tubular vertical structure

RV "B" Craft #1 w/energy beam

Craft # 2

Impressions from Remote Viewer **"B"**:

- The ship was… "Circling overhead."
- *"I felt like the witness was familiar with the ship."*

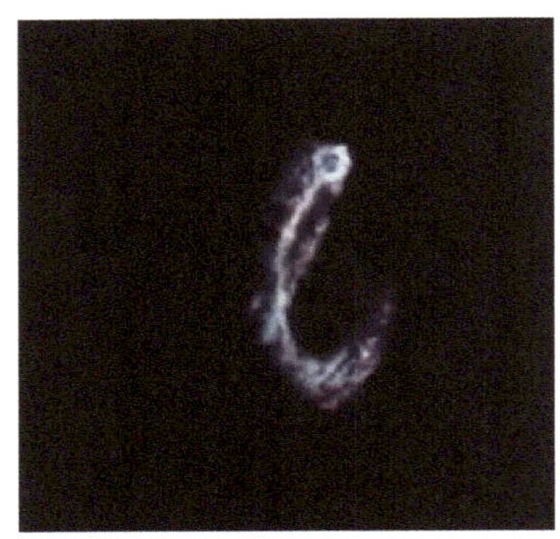

Ship circling

Commonalities:

- Large pole or tubular structure.
- Energy coming out of pole, or from craft.
- Green (grass) surrounding the pole
- A rounded rectangular structure nearly identical to the photo, (white or tan building with roof)
- A craft had circled and landed. Witness was familiar with ship.
- Craft was circular or disc-like per RV drawings, and photo.
- Two viewers described a band around the perimeter of the disc.

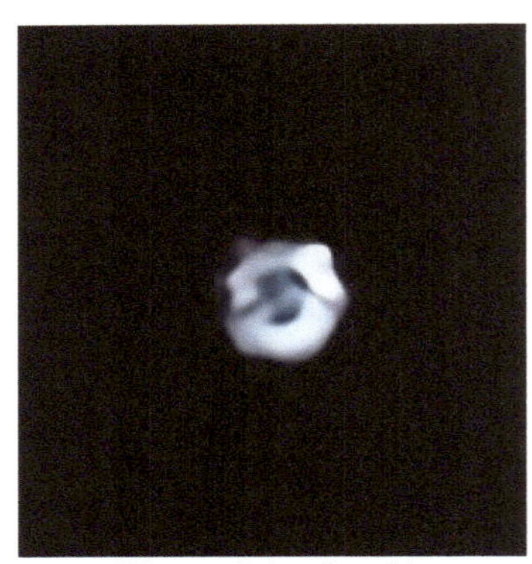

Photo of circular ship

Remote Viewing

RV Session 7
Date: 06/24/2021
Case # 115969

Stepped Roof

On June 24, 2021, four remote viewers were given the case number and no other information and asked to view the event and circumstances.

The Witness in the case stated:

- He saw several entities, two of which were in black robes.
- One of the entities seemed to be bending down.
- Grass and trees are beyond the carport where the entities were situated.

The witness' video shows:

- Entities in black robes with big round eyes.
- **Several entities.**
- Park with trees and grass beyond the carport.
- **Stepped** roof of the carport.
- **Two rectangular roofs.**

My Impressions:

- I initially saw **2 square or rectangular roofs**...
- One roof seemed lower than the other and was smaller. They were **stepped**.
- I also saw **greenery**.
- I had an impression of round eyes.
- I saw **several entities**, whose heads looked somewhat normal but with **round eyes**.

My drawing of the stepped roof

Right: Witness' photo of stepped roof

<u>Commonalities of RV impressions, and case data</u>:

- Stepped roof drawing like the video
- Two rectangular roofs
- Greenery in area is from grass/trees beyond carport
- Round eyes drawing is like video of entities
- Witness stated there were **2 black-robed** entities.
- Several figures have round eyes.

Drawing of eyes

Right: Photograph of entities with black round eyes or goggles.

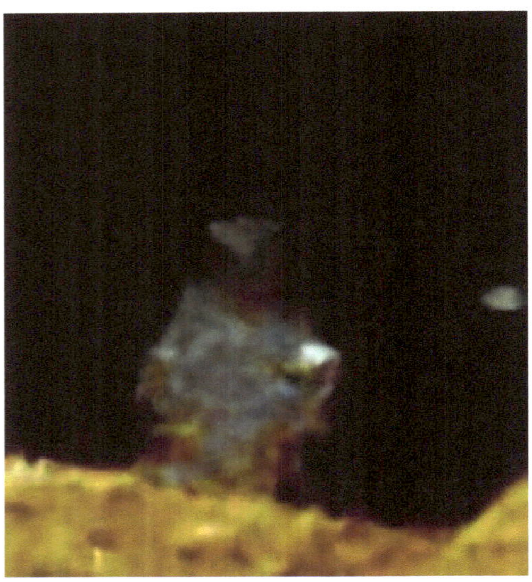

Left: Frame from video of NHI with round eyes also

<u>SUMMARY</u>:

There are six points of commonality between the RV impressions that I had compared to the report from the witness and his video images.

The drawings are remarkably similar to the items in the video.

Remote Viewing

RV Session 8
Date: 06/13/2021
Case # 116418

Several Round Objects

Remote viewers were given the case number and no other information.

My Impressions:

A. I saw an oval disc in the sky. The witness was outside the house.
B. It was not dark outside, but near the end of the day.
C. The disc seemed to shoot out a light below it. Then I saw another, maybe 2 discs though at the time I thought it was the same one.
D. I first saw a gold-colored ornate roof line when I began to view the case.
E. I saw an extraterrestrial with an elongated face, and eyes different from humans.

other

The witness stated:

A. The witness saw round objects in the sky. He was outside the house.
B. The time was 5:13 pm.
C. The witness saw several round objects. Videos also show at least 2 white round objects and 3 dark objects.
D. A gold-colored ornate roof line is present 3 houses down the street from the witness' home.

Drawing of object

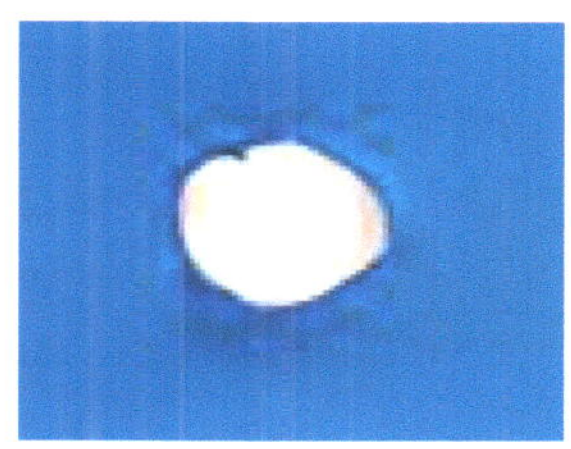

Frame 15:11 from video

6 POINTS of COMMONALITY of the RV Session:

1. Round object somewhat oval.
2. Round object in the sky.
3. Witness was outside.
4. End of day, still light outside.
5. Over 3 round objects.
6. Gold colored ornate roof nearby.

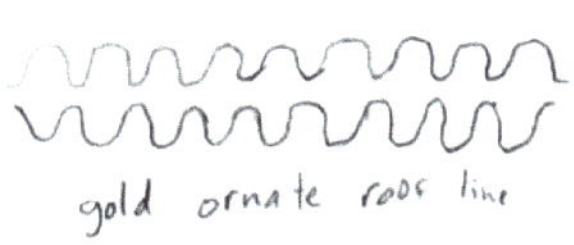

Similar gold ornate design

Remote Viewing

Front of a House

RV Session 14
Date: 03/08/2022
1136 S. Pearl

On March 8, 2022, four remote viewers were asked to view an address, with no other information. Here are the results from my remote session:

- I saw features of the front of the house including pillars, windows, roof, iron railing, front door, steps to the porch, sidewalk, etc...
- I saw what looked like a fireplace inside the house which may date to an earlier time.
- I saw a grate or panel though I was initially unsure where it was.
- There was also a very bright white light…in a darkened area. I thought perhaps it might be in a basement, but it could have been somewhere else. The light seemed small. I got the impression that it was related to a ghost or spirit which was visiting from time to time and going to other houses too.

ITEMS CORRECTLY IDENTIFIED:

1. 2 Pillars.
2. Both pillars supporting a curved piece.
3. Half-moon window in front door.
4. Windows on both sides of the door.
5. Iron rail/panel joining pillars to house.
6. Slanted roof over pillars and door.
7. 3 steps down to the front lawn.
8. Sidewalk going straight from the steps.
9. Impression that a ghost visits from time to time.
10. Impression that ghost/spirit visits other houses too.

Small bright light/spirit

Note: *The owner of the building, who has an office in the house, indicates that for over 23 years, the spirit of an old lady has visited the house and caused assorted items to disappear, and then be returned sometime later. She has also been seen multiple times by office staff. The owner says that the spirit also visits other nearby houses.*

Remote Viewing

A Brilliant Sphere – Viewed Remotely

RV Session 1
Date: 03/21/2023
Brilliant Sphere

*While adding the case in Chapter 2 regarding **"A Brilliant Sphere"**, I wondered if there was anything else that could be gleaned from that occurrence. So, I decided to remote view it from the perspective of the sphere. The sighting originally occurred on Aug. 31, 2014. My remote viewing was on March 21, 2023.*

Impressions from the perspective of the sphere:

- As the sphere hovered above the ground, a house in the distance was lighted up by a small bright light in a window, and it (NHI) knew that someone was observing the sphere **and recording it**.
- When the sphere turned to the west down an uninhabited corridor, there was a decision to turn a red light on behind the sphere as a subterfuge to make any other observers think it was a normal aircraft, or less unusual to potential other viewers.

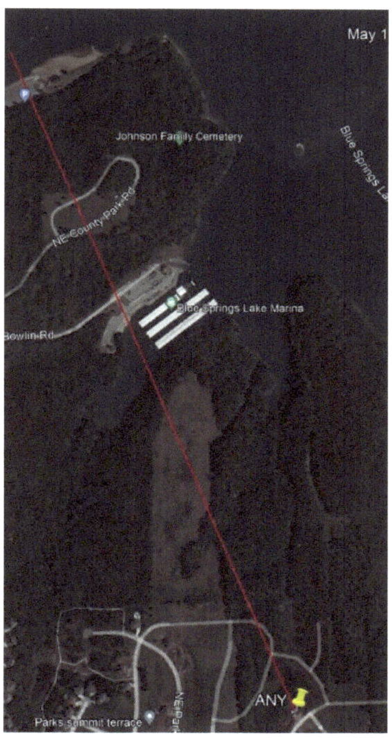

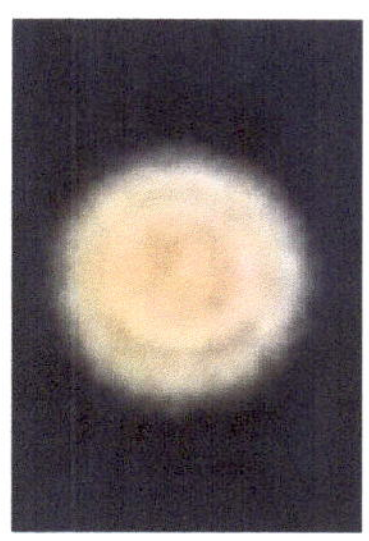
Sphere

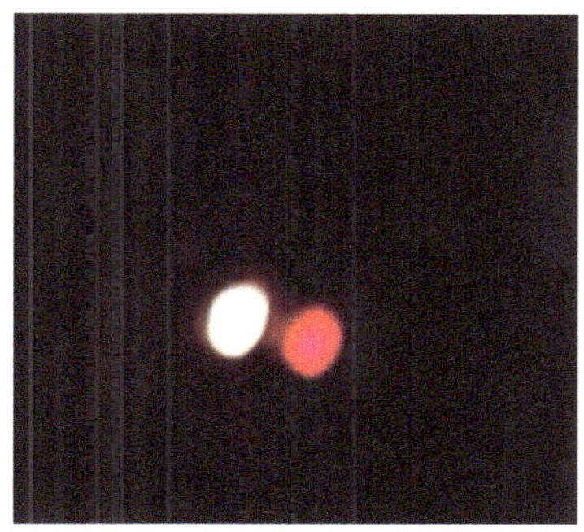
Sphere after it "morphed" into 2

Line of sight view from the sphere toward the house where it was being recorded (yellow push pin).

> **NOTE:**
> This remote viewing session was unlike others because I sensed the thinking of the NHI driving it:
> 1. I could tell that it realized that it was being viewed and recorded.
> 2. I also noted the decision to make a red light appear as it left the area.

Remote Viewing
Four Glowing Craft

Case # 108984
Liberty, MO
May 16, 2020

This image-- # 2237-- was taken 43 seconds after image 2236, and 48 seconds prior to 2238.

This case was filed by a journalist who was photographing his classic car (1964 Ford Galaxie) on the square in Liberty, Missouri. Coincidentally, it is the same car that was used in filming Men in Black III. He took eight photos. Later, when he looked at the images, he realized that there were four to five objects that he had not seen with his eyes in one of the photos. They obviously had been just caught by the camera in a fraction of a second. He did not notice them in other photos. If they had lingered in the sky, they most likely would have been seen.

The reporter was asked to return to the location of the sighting and open the SkyEye App, then take a screenshot of the view of the sky and associated stars and constellations. Knowing the closest star, and after consulting star charts at that time of day, one can know the angle of elevation of the star. This is one way to know the angle of observation during the sighting.

This image was taken during daylight hours six days later.

The angle of observation was 23.5 degrees to the largest and brightest of the objects. Since the cloud ceiling was at 1363 ft, and the final object at the far right appears to be coming out of the clouds, the cloud bas is a good indicator of the altitude of the craft. The base distance from the witness to directly beneath the objects was .59 miles.. This allows us to calculate the size of the brightest craft to be about 30.5 feet.

I had asked Margie Kay to remote view the four craft to see what additional information she could add… on 11/3/2023 she answered:

*"I'm seeing silver disk-shaped objects that are approximately **30-35 feet** in width with **domes on top**. The sunlight washed them out. Looking inside the craft, I see tall greys at controls, most are sitting, one is standing and walking around. They are communicating telepathically.*

*They are on a mission and will be landing at a site nearby but in a farmer's field. They will cloak when they are on the ground. I see samples being taken of the ground and vegetation, and they are interested in an animal nearby. It is a small horse I believe. They send out two beams of some type of energy, which pulls the animal towards their craft. I have **never seen** this before.*

The animal is taken into a cargo area on a ramp that comes off the craft. It is put into a trance-state, and something is injected into its neck.

There is a pond nearby and I can see the farmhouse about 300 yards away. There is a fence around the property and tall weeds on the other side of the fence where the farmhouse is located.

If I get anything else I'll let you know, but these were definitely craft."

On 11/4/2023, I asked Margie further questions about the greys that she had RV'd…

"I saw tall greys – maybe 5' tall or so sitting at a console in the far-left disk, which seemed to be the lead craft. I saw 5 sitting at consoles, and one standing watching and walking around.

They did not seem affected by speed or turns of the ship. I know there were beings in the other craft but did not RV them at the time. I'm looking now...

*A very tall being with a large, rounded head is in the second craft in the center.
It is about 12' tall and is unlike the greys. It is a tan color with a protruding forehead.*

The rest of the craft have the same number and type of beings as the first, with one standing and 5 sitting, except in the last one, there are only 3 at a console."

Of Note:

1. Photo analysis found the brighter craft to be 30.5 ft in diameter, with a dome on top.
2. Remote viewing found the craft to be 30 to 35 ft in diameter, with a dome on top.

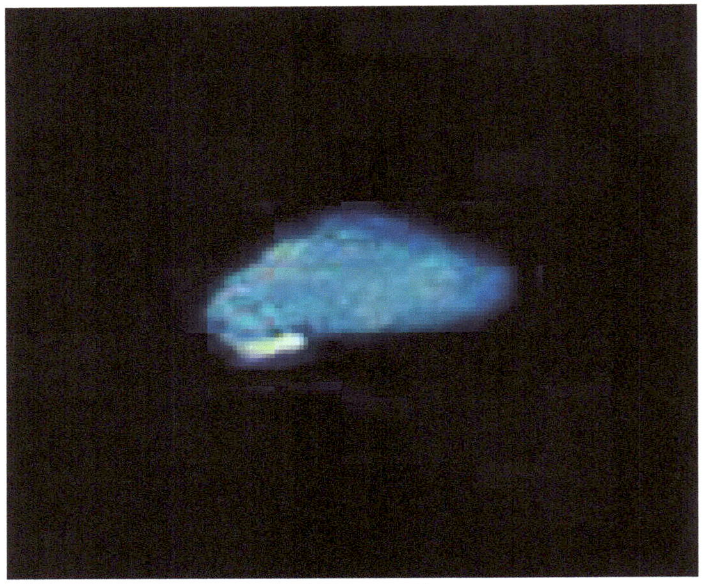

Brightest craft

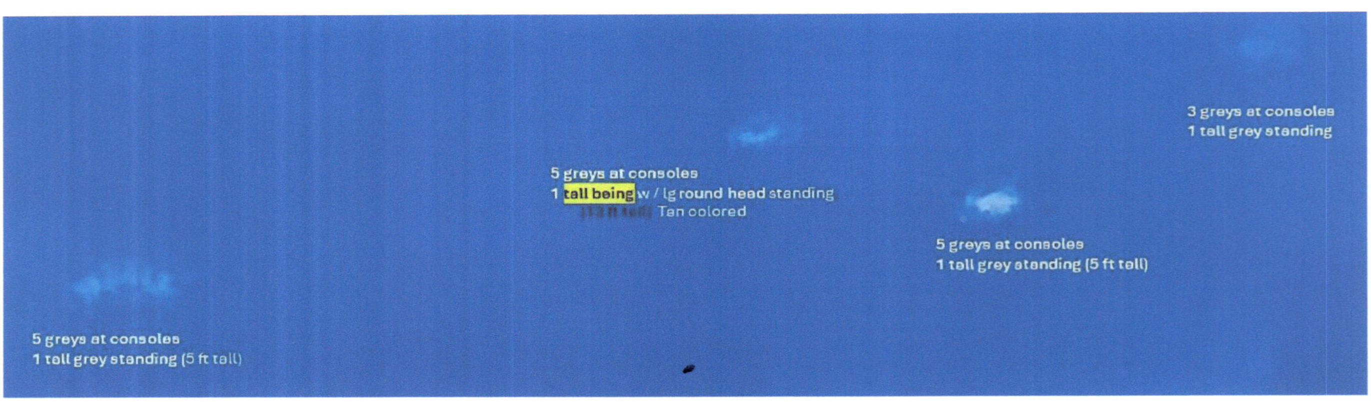

Detail provided by Margie's RV

Remote Viewing

Extra RV Session:
Lily Nova CE5
(May 6, 2024, RV Session)

On the next page is a summary of three Remote Viewers' notes regarding a CE5 experience by Lily Nova. She had gone to Lakeside Park, in St. Peters, MO to meditate and invite any close-by UAP to show themselves. The first four photos of the craft were taken automatically by Lily's camera. The Remote Viewers were asked to view and describe the event.

The following sequence concerns 4 photos and data concerning them from Remote Viewing:

Photo 1: *(DSC_313).*
At 6:16 pm, the first photo was taken by the camera.

Photo 2: *(DSC_314).*
From 6:16 pm until 6:28 pm, the craft rapidly descended, and the entities inspected Lily. This second photo was recorded by the camera**, but was removed** by the entities due to too much detail.

Photo 3: *(DSC_315).*
At 6:28:39 pm, the ship had moved back into the sky and the third photo was taken by the camera.

Photo 4: *(DSC_316).* At 6:28:44 pm, (5 seconds later) the 4th photo was taken by the camera.

The image is from 3 stacked images. Lily is meditating. The arrows show the movement of the object.

CE5 3rd image

Compilation of Notes from 3 Remote Viewers (5/6/24):

- Lily drew the craft to her…the entities were curious, noticing Lily's light which had expanded out quite some distance.
- The craft was black and silver, shifting color back and forth. After the first photo, it moved very rapidly downward at a 20–30-degree angle of descent.
- The ship altered time, traveling in a time dilation funnel as it moved toward Lily.

- Inside were several grey entities with long arms, grey suits, large heads and eyes, grey boots and hats (like Navy dress hats). It was sensed that they're extremely high vibration beings.
- After the first photo was taken, the craft moved down toward Lily.
- While in the time funnel, the craft dropped out of 3-D as it was moving.
- The entities came in remarkably close to her (about 19" away) to scan her brain to see what level she is at right now.
- Twelve minutes elapsed as the entities inspected Lily, and then the craft rose back up in the sky where two more photos were captured in rapid succession with an interval of 5 seconds between the two images.

- The camera was allowed to record four photos, but the entities removed the second image (DSC_314) because it was taken while the craft was next to Lily and contained too much detail in the photo.
- After the third photo, they pulled back from our dimension.
- The entities communicated by telepathy.
- The entities know that we remote-viewed them.

NOTE:
The first photo in the series taken by Lily's camera bears an odd timestamp: **—June 4, 2014, at 5:43 pm— 7 years earlier!!** All other photos timestamped about **94 minutes earlier than the actual time.**

Remote Viewing

Overview—

I have personally been involved in Remote Viewing for **seven** years (*as of July 2026*). It provides a unique way to access information in ways that we do not grow up knowing about.

As we develop from infants to adults, we are socialized into a world that is "nuts and bolts" oriented. Our science and accessible technology are grounded in Cartesian and early Einsteinian physics. Our aircraft use airfoils to provide lift, our vehicles still mostly burn petroleum fuels, our steel mills heat metal and pound it, roll it, perforate it using mechanical methods. We are using computers today that are amazing in what they have developed into over a period of about 55 years. Satellites link us, underwater cables connect us, and wireless services abound.

But all of it is mainly "stuff." Bits and pieces of metals, crystals, wires, and electricity. Remote viewing is something that is theoretically available to most persons given the proper environment, training, and interest. And it is still a field pioneered by just a few individuals. Yet it is very real.

Summary:

I learned how to view remotely from Margie Kay. Remote Viewing has shown me that there is another aspect to the world and to communication that is a part of our humanity…just unexpressed in the general population. The cases depicted in this chapter show how important this form of communication can be.

What Does This All Mean?

The three preceding chapters have covered 23 cases in which several types of Quantum Communication took place.

In **Chapter 1**, (the Fast Movers Update), we looked at **8 cases**. Six cases involved intention to find UAPs and two involved accidental discovery of UAP.

In **Chapter 2**, (Quantum Communication: Conscious Intention), all **7 cases** involved NHI showing itself. Another 3 cases involved contacting the NHI as well. In 1 case the witness was unsure why it was recorded, and in another a UAP was discovered accidentally.

Chapter 3, (Remote Viewing) All 8 sessions demonstrated how remote viewing can access information and data that are perfectly accurate. In some cases, the information was corroborated between remote viewers. In other cases, the viewers' information could be corroborated with original witness information. Two sessions also involved telepathically contacting the NHI.

Of the sessions which involved four remote viewers, a total of 48 common points were corroborated by remote viewers, and/or witness information. That is an average of 8 points of agreement per session.

All six types of quantum communication were involved in the 23 cases which have been presented. They include:

1. Contacting NHI
2. Communicated intention
3. NHI showed itself
4. Remote viewing
5. Unsure why recorded
6. UAP discovered accidentally

I continue to be amazed with each new discovery, and to the continued expansion of the universe which unfolds as I learn, am shown and taught, and experience . This is like discovering a new organism that permeates the world, which has many roots along the way. It branches into all our known human structures, societies, and links itself to other worlds, galaxies, and dimensions (universes).

According to some, we never were just one humanoid species on planet Earth.
This book provides evidence of "others." This of course continues the insights that we discovered and shared in the first book entitled "***The Fast Movers***." Along with multitudes of other books written by many authors and experiencers who witness of other types of craft, this book testifies to communication elements that are involved in these extremely fast moving UAPs and other types.

Earth's Unseen Inhabitants allows us to see some glimpses about those who are sharing our space on planet Earth. Most of the time they are unseen in their physical forms. The evidence primarily is photographic and experiential. This has been the purpose of this book...to show yet again that there are many others that surround us. And there are a lot! We know that many other persons have had direct contact with Non-Human Intelligence in many forms...including abductions. They are to be found in many other works including books, video presentations, seminars, in a variety of media including documentaries, whistleblower accounts, and congressional investigations.

More can be said by those who have other experiences with NHI…have communicated with them, been taken onto their craft and shown wonders of our solar system and beyond. But that is another story.

"There is no doubt in my mind, after 37 years of study and investigation that the evidence is overwhelming that planet Earth is being visited by intelligently controlled vehicles whose origin is extraterrestrial. There are no acceptable arguments against flying saucer reality, only people who either haven't studied the relevant data or have a strong will not to believe that Earth is at the bottom of the heap sociologically and technologically in our local galactic neighborhood."

Stanton T. Friedman

Space Photography, ET Contact & A Great Awakening
by Lily Nova

Ever since I can remember I was a very intuitive individual. I seemed to understand things at a deeper level and viewed things through a different lens than most of my peers. I always felt drawn to space and the mysteries of our history and the universe. Where did we come from? What happens after death? What else is out there waiting to be discovered in the vast universe? What is life? These are the types of questions I often pondered. That being said, nothing could have prepared me for what I would come to experience and discover during COVID.

My 'BIG awakening' as I call it would begin in the year 2020, in the middle of the COVID pandemic. I was a nutritionist at the time and taught nutrition and cooking classes around the St. Louis community for Mizzou University Extension. After an abrupt series of unfortunate conflicts with my boss in September 2020, I quit the 'dream job' that I worked so hard to get. I was heartbroken and felt lost.

In the midst of pandemic lockdowns and freshly unemployed, I turned to the stars. Space photography was a hobby I picked up during the lockdowns, and now with more time on my hands, I began spending every single night out under the stars. It was the only place that felt like home and gave me comfort in uncertain times. I focused all my attention looking up and became exceptionally good at photographing the milky way, the moon, and deep space objects. I fell madly in love with the night sky.

My photo of the Andromeda Galaxy

My photo of The Pleiades Star Cluster

After spending months nonstop out under the stars, I had my first UFO / UAP experience. On November 17th, 2020, I went outside at 1:50am to get some fresh air before bed. I stepped out on my front porch in the suburbs and looked up at the sky like I always do, when I instantly locked eyes with a strange moving light. It was larger and brighter than any star or planet and couldn't have been a normal aircraft because it was moving back and forth in a circular motion, hovering in the same spot above the neighborhood.

My eyes widened as I questioned if I was actually seeing a UFO and I grabbed my phone to begin recording it. It was so dark my phone camera was going in and out of focus. A strange blood-orange light suddenly appeared down the street that drew my attention. I looked over to investigate it. Suddenly, time froze, and I became hyper-aware of how bone-chillingly silent and empty it was outside. I looked around at a brisk autumn night, leafless trees stood perfectly still and silent as dark silhouettes against a clear star-lit sky, with no wind and nobody else around. It was so quiet you could hear a pin drop.

Suddenly, the wind began to rustle, and I looked back up to find a much larger and closer second craft had appeared seemingly out of thin air! It was a strange diamond-shape, dark gray with a row of red and blue-green lights that went across the middle. The craft was maybe a few houses down as I stood there bending my neck looking up at it. I was in shock and awe… I was the only one around and I knew this craft was looking right at me. The diamond UFO darted to the left, then to the right, then shot up and back down at insane speed to show me that it was not a normal aircraft. Then the craft slowly started coming straight towards me. I began to tremble in shock, not sure what to do. The craft grew closer, and I decided to step back under my porch and hide under the roof as it went right above my head. After a few short seconds went by and I realized it wasn't going to drop a beam of light and pick me up, I jumped out from under my porch to see where it went. It was gone, completely vanished!

I intuitively knew that the UFO must have either flown straight up at ridiculous speed, or it popped into another dimension where I could not see it. I was shaken up and didn't know what to think or do. 'They' made it very obvious they saw me and homed in on me. My whole paradigm instantly shifted, and my life forever changed. I could not sleep that night. That was the first encounter of many.

The UFO's gave me a couple of months to integrate my first encounter. In January 2021 I would have my second encounter while alone at a local astronomy park. At around 1:50 am again while sitting in my car, I would suddenly become surrounded by strange lights in every direction, including about 50 yards in front of me and my parked car. I became very curious, unsure if the UFO's knew I was there since my car was not started and I was the only one around surrounded by corn fields (I would later discover they knew I was there, and they were there for me).

I began filming the lights with my Nikon D5300 DSLR camera. I did not want to start my car and draw attention to myself, so I decided to record them and wait them out. Hours passed by as I watched the strange lights behaving very strangely, unlike anything I had seen before. They were not leaving, and they acted stranger as the night went on! I filmed them until the sun began to rise and I saw another car drive down the road, then I finally felt comfortable enough to leave the park. I was completely mind blown at what had just happened, and that I was able to capture much of it on film. I literally spent the entire night surrounded by UFO's.

I would return to the astronomy park alone the following night to see if they would show up again. I did see 2 of them hovering back and forth above the tree line on the horizon, which I captured on film. I went back the third night with a friend, determined to have somebody else witness this. That night they did not reveal themselves openly, but something strange flew past me when I separated from my friend to walk to the car. They were letting me know they were there. Even though my friend did not see them, he could feel their presence and said he felt like we were being watched. When I returned home that night, there was a light in the sky waiting for me. It slowly moved around, and I recorded it and spoke to it until the sun began to rise, then I went to bed.

I would see about 13 UFOs / strange phenomena within the next 2 weeks and caught many of them on film. They seemed to be jumping in front of my camera, like they wanted to be captured. They also seemed to be trying to tell me something.

Beam of light comes from a craft in front of me during my 2nd encounter at the astronomy park

With so many encounters occurring suddenly, I knew something big was happening. I found MUFON while researching and contacted them to report my experiences. They seemed to be shocked by how many close encounters I had in such a short span of time, calling my case 'very peculiar.' MUFON asked me what I had been doing that may have brought on these experiences - my response was spending time under the stars, and I had also been studying the science of consciousness. I would later realize how pivotal consciousness is in all of this, how having foundational understanding of it played a huge role in my ability and openness to have these experiences, and how pivotal it is in deliberately making contact with extraterrestrial intelligences.

I intuitively began speaking to the beings who were controlling the craft I was seeing, unsure they could hear me at first. I asked them why they were appearing to me and if they wanted me to share the footage of them, to show people they are here. I quickly began receiving responses to my questions. I was completely amazed, and mind blown that intergalactic beings not from Earth were appearing and communicating with me. I wanted to know everything. I was determined to get answers, capture evidence, and share the truth with others.

I began meeting with these beings daily. I went out with my camera every day, reaching out to make contact (often called 'Ce5' or human-initiated contact with ETs / UFOs). At first I went out at night, but after experimenting and capturing my first orbs of light during the day, I switched to daytime and sunset. We made an agreement and established a regular daily routine. I found that when I would get in a 'flow state,' they would appear. I would go out with the intention to make contact and would begin filming the sunset or clouds, getting in that blissful, creative flow state, and then 'boom!' I would feel a rush of energy and love when they were around. The almost overwhelming feelings of positivity, excitement, and love I felt during these contact sessions… is hard to even put into words. It is wonderful.

Me meditating during a Ce5 contact session with a ship to the top right

Close up of the UFO/UAP

Close up of two black triangles and what appears to be a portal opening

As my relationship with the extraterrestrial intelligences developed and our meetings continued, my psychic abilities were being activated rapidly. I began to receive empathic, intuitive, clairvoyant and telepathic information. I was receiving what I would come to know as 'downloads,' or packets of information, activations, and wisdom. I was not psychic before the UFO encounters, this was activated by them and rapidly strengthened by my dedication to continue meeting with them and training daily. They instructed me to begin meditating, which would pave the way to countless profound experiences.

When the extraterrestrial intelligences first began communicating with me through telepathic vision, I got to see what they looked like and learn where they were from. The first beings I saw happened during one of my first Ce5's when a golden orb of light appeared (see photo). I closed my eyes and received telepathic visions of the crew, who had light blue skin, silver skintight uniforms on, and no hair. They later told me they were originally from the Lyra constellation, then migrated to the Andromeda constellation.

I also discovered this was my Star Family. I discovered that I was a Starseed (I was one of them - an old soul that did not originate from Earth) and was on an important mission to help humanity through disclosure and a great awakening. As I progressed, I would see many different races of beings, including beings from the Pleiades, Orion, Sirius, Arcturus and others. They all teach me things about who we are, who I am, how the universe works, and they help me unlock higher levels of consciousness and activate new gifts and abilities. Contact with these beings has been an incredibly profound and enlightening experience. Now I share my experiences and teach others how to make contact too.

Golden Orb of Light, The Andromedans

Here are some things I've learned from meaningful contact with my Star Family:

We are all connected through consciousness, and you can make telepathic / psychic contact with these beings no matter the physical distance between you. Everything is frequency and vibration. Many of the extraterrestrials interacting with Earth are extremely high vibrational / in a higher density than 3D, and in order to perceive and connect with them, we must raise our frequency to meet them. Humanity is unaware of how powerful they are, we have been suppressed and lied to about everything. We must remember who we truly are, and begin activating our dormant DNA, unique gifts and raising our consciousness. We must also focus on healing and releasing our old baggage. It cannot come where we are going (the New Earth).

We all have a Star Family who cares for us very deeply and is watching over us (especially if you are reading this). You can connect with your Star Family, too. Set the intention to make meaningful contact with benevolent beings, spend time outside looking up, raise your frequency or get in a blissful, loving state or creative flow state, and develop a regular meditation practice.

Section 3

Unidentified Aerial Phenomena Answered

by Bill Spicer

The time of awakening, as I have come to call it, began in my teen years with a UFO/UAP sighting at night while riding as a passenger in a friend's car, while on a rural highway. There was no premonitions or warning this was going to happen, but a bright flash of light illuminated the car, and a red and green bright set of lights fell out of the sky in front of the car, blocking the roadway as we came to a complete stop. I watched what seemed to be a few minutes then it just disappeared moving quickly out of sight into the night sky. The others who had been in the car with me when questioned the next day did not recall any of this experience. I thought, *how odd, or could it have been an aircraft that lost its way*? That same day I visited the local FAA Flight Service Office and asked if any aircraft had been reported missing. Of course, the answer was no. This began my search for answers.

As a young adult while working outdoors, I was stung by bumble bees and required immediate medical attention due to a severe allergic reaction that caused anaphylactic shock. I stopped breathing. During this process I had a near death experience. Totally went out of body for a what seemed like 15 minutes in a dark dream like state like waiting on something to happen, before being pushed back by an unseen force, into the body while in the emergency room, was told I had more to do. After this experience the ability to meditate and leave the body, to become astral began, along with very lucid dreams that became precursors, or precognition to events in my life before they unfolded. Through discovery I learned also how to remote view when meditating as an out of body experience, from which I would later go on board the craft after seeing them using Quantum UFO Observation Technique.

Starting in 2009, the final pieces of the puzzle began to fit together. I started having lucid dreams where I was taken aboard a large multi-story UFO/UAP craft and was being guided around by a tall bright white light emitting robed figure that appeared to have a higher level of thinking or consciousness. There was no spoken communication, but only mental telepathy. During the first of three such encounters, I was told first that I would begin to see UFOs/UAPs and then how this was to be done. It involves meditation, by taking time to ask the higher consciousness beings to show their craft, then using a digital camera and a polarization lens to filter the image. This process is what I called Quantum UFO Observation Technique.

Through additional lucid dreams I was shown how the UFO/UAP beings use mental imagery, or visual thought projections – consciousness, quantum electrodynamics – inflationary vacuum state propulsion to propel the UFOs/UAPs and can go from matter to energy and cross space/time using quantum physics, based on non-locality – Space A equals Space B. There are examples where I have video and still photos showing this process of coming and going through worm holes – event horizons or leaving through what appear as doughnut holes through the fabric of space/time that open up instantaneously then close.

In 2010 I authored a book of my experiences about being able to see UFOs/UAPs during the daytime using Quantum UFO Observation Technique, by filtering (linear polarization), to record the UFOs/UAPs using a digital camera. I understood very well by this time that there was a quantum event taking place each time I went out to view UFOs/UAPs. I would also receive precognition knowledge on when to start the meditation and go outside to observe and have concluded that consciousness is the driving force behind everything.

I have completed a detailed, step by step method on how to raise your level of consciousness and see the UFOs/UAPs. This technique is a gift to help awaken humanity and raise the level of consciousness to see the higher consciousness beings craft as they move about extremely fast in the space/time around us.

The Unidentified Aerial Phenomena is answered by evidence of a Quantum Consciousness Field made up of quantum electrodynamics (QED) where the relativistic quantum field theory of electrodynamics takes place and human consciousness meet.

By using this viewing technique, I have assisted others with the awakening process so that they may understand their experiences of contact with a non-human intelligence, and related paranormal contact experiences. The contact, "Contact Modalities," may include contact by: Near Death Experiences (NDEs), Out of Body Experiences (OBEs), UFO/UAP visitations, Meditation, Remote Viewing, Lucid Dreams, etc.. These experiences form the modern miracles.

The Unidentified Aerial Phenomena (UAP) discussion is a place to start telling humanity about the truth of the existence of higher consciousness beings interacting with humans here on Earth. I wish to leave the reader feeling there is hope and knowing there are higher consciousness beings watching us as we raise our consciousness vibration and to project a thought of living in harmony. As the public becomes more aware and is starting to awaken to the knowledge that higher consciousness beings are here as a precursor to something bigger that is coming to humanity.

Included are recent examples of UFOs/UAPs I have seen making appearances using a cellphone camera with polarized lens filtering and examples of what are known as lens flares, often seen when aiming a camera towards the direction of the sun.

Recent UFO/UAP Examples – Video Analysis

August 16, 2024

There were two UFOs/UAPs seen in the upper center part of the video which varied a lot from frame to frame, with one becoming smaller and eventually disappearing. The size of the UFOs/UAPs was small and at a distance. What was seen consistently was a black rod-shaped UFO/UAP. The one thing that was occurring is that the UFOs/UAPs were high enough to be hitting some low-level clouds and disappearing, then reappearing as they hit a patch of sky that was clear.

Frame 1:23 enlarged

One version of the UFO/UAP is in Frame 0:02 shown below.

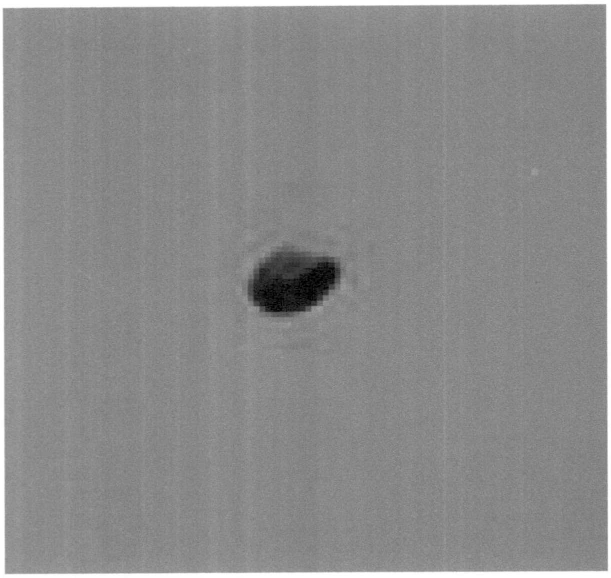

Frame 0:02

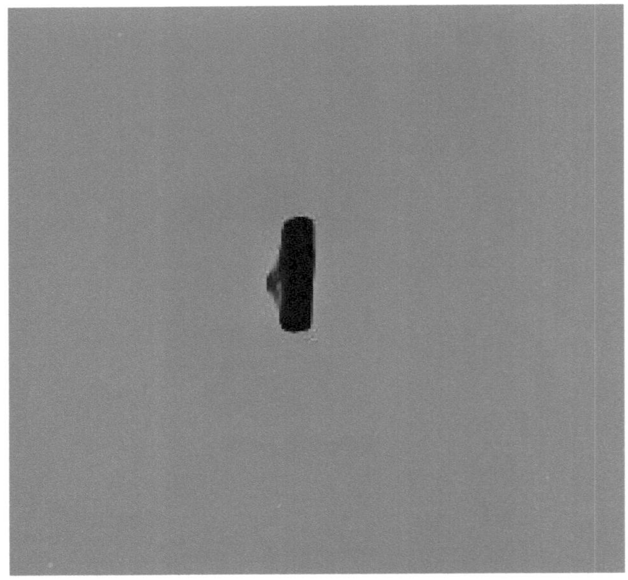

Frame 0:26

The constant shape in the frames is the black rod-shaped UFO/UAP. The UFO/UAP was found to be at an altitude of 5, 294 feet altitude, with an angle of elevation of approximately 30 degrees, and estimated to be 1.8 miles away from the camera. The UFO/UAP has been calculated to measure approximately between 6 to 12 feet in length.

Enlargement of rod-shaped UFO/UAP:

UAP Observed on July 8, 2025:

First Enhancement Frame: 2.22

Original frame

Enlargement of UFO/UAP

First enhancement

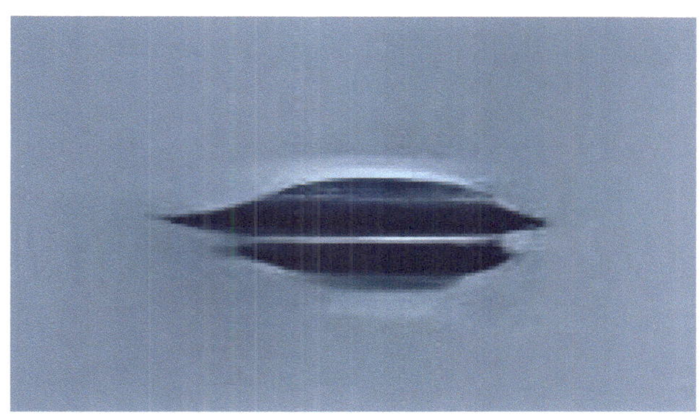

Second enhancement

Third enhancement showing detail of craft

Fast-Mover Observed on January 18, 2025

Right: Original photo:

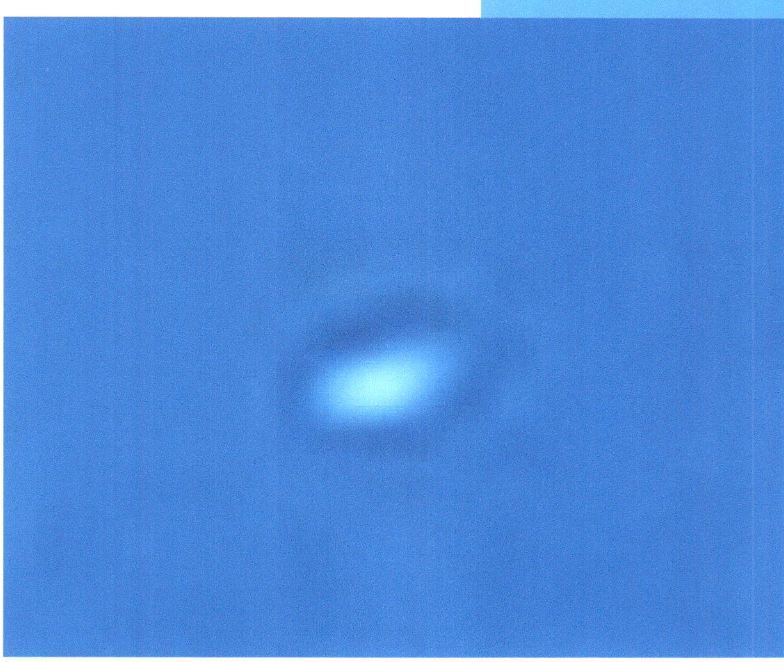

Enlargement of UFO/UAP

UFO Observed on June 29, 2025

Enlargement of UFO/UAP

Another Fast Mover UFO Observed

Enlargement of UFO/UAP

IR Night Vision Photos Taken July 13 and 14, 2026

Links to YouTube videos I posted recently showing the movement. They move in our skies.

https://youtu.be/w9dpzqNcEcc?si=X3eeHi-f1coKcFZ0
https://youtu.be/m36Xxb4D7JI?si=kH3iitIioeDQTKlo

Detail of craft using new IR Night Vision - These new sightings have awakened some of my close friends. I still use the meditation before going outside and telepathically communicate asking them to show themselves.

I point the camera towards the North Star and set up using 8X zoom - 1080p, camera tripod and full IR setting. Also have caught Star Link satellites moving from North to South, but they are not like the objects I am seeing. These objects just pop up and start moving!

See videos of these objects on YouTube:

https://youtu.be/w9dpzqNcEcc?si=X3eeHi-f1coKcFZ0

https://youtu.be/m36Xxb4D7JI?si=kH3iitIioeDQTKlo

Lens Flare Photo Analysis – Examples

Original Photo

SUN

Sun W/Luminance Gradient Analysis

Example 1: The study with Tone 4x is an exact duplicate of a structure in the base of the image of the Sun that was run through Luminance Gradient Analysis. This means that the object is: A lens flare and nothing else.

Figure 8Tone

Figure 6Tone 4x

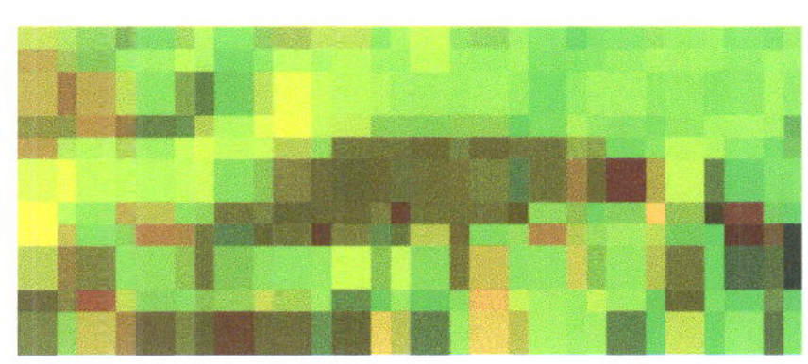

Figure 7Sun Gradient

This analysis demonstrates how a part of the sun that is visible can create a lens flare of a different size. The streak above the white oval lens flare is also another type of lens flare.

Example 2:

Left Image

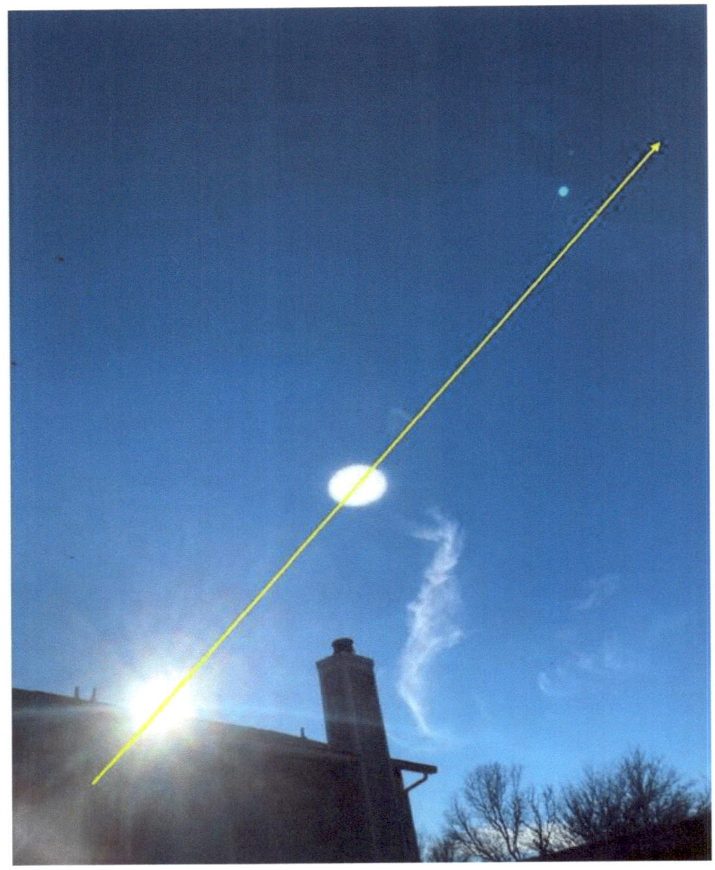

Right Image

Solar Flare Examples

The oval shape in the left image is virtually identical to the image on the right. The relationship between the sun and the oval Lens Flare is--- they are the same. The sun causes camera lenses to reflect light internally which shows up in the image.

The right image is an example of **<u>three</u>** <u>separate lens effects</u> from different parts of the camera, causing slight variations of the dispersement.

Solar Flare Example 3

This is another example of lens flares. You will notice in this case the perfect alignment of the faded white oval, and the turquoise round dot, compared to the Horizontal and Vertical lines. This is a classic case of lens flaring.

- This technique with the horizontal and vertical lines works primarily with a full photo…not one that has been cropped.
- There are also many other kinds of lens flares… Out of frame influences, astronomical, and others…

Skeptics sometimes dismiss video and photos of fast movers as lens flares, however, all the photos in this book have been analyzed and lens flares have been ruled out.

"I believe there's other forms of intelligence in the universe. I've seen and heard some pretty convincing UFO stuff. Besides, if we're the most intelligent things in the universe... well, that's just depressing."

Rekha Sharma

Section 4
Communication with Non-Human Intelligences
By Margie Kay

By the time I was three years of age I knew that I had the ability to do things that others around me did not. I saw energy fields around people, animals, and trees. I knew when a certain person was about to call on the phone, or if an accident had occurred or someone was lost. At first, these things were not something I paid much attention to, but my family did not discourage my questions or abilities even though they did not possess the same. Both of my parents were well-educated and highly intelligent. My mother had a BA degree in Business and worked as the Credit Manager for a large corporation. My father read over 10,000 books and spoke seven languages. He taught at the University of Southern Missouri even though he didn't have a degree in any subject. I believe that my parents were smart enough to let me develop my psychic abilities on my own without interference or discouragement.

As a child and throughout my life I received assistance from an unknown voice who would speak to me telepathically. I heard his voice only inside my head. He never spoke more than a few words, which would just be enough to convey a point. What is odd is that I hardly noticed the voice at all as a child, and the ideas would simply come to me. I didn't realize until much later in life that this had been going on the entire time. *Note: For the remainder of this chapter I will refer to this entity as "The Voice."*

I started out as most people do who are psychics and remote-viewers– being psychic as a child and wanting answers and more control over my abilities, I did research on the subject. I also practiced these skills every day for most of my life since the age of 11. This constant practice is the reason that I have been able to activate the psychic center of my brain, which eventually led to contact with inter-dimensional intelligences.

Before going further on this subject, I think it is important that the reader understand my history which eventually led to my direct contact with Inter-dimensional beings. I had no formal training, though starting at age 11, I started doing research to learn more about how clairvoyance, clairaudience, and clairsentience works. My mother took me and my brother and sister to the library every month, and I always left with the maximum allowable books about psychic phenomena, science, the occult - anything to understand the sixth sense world - and devouring each book before our next library visit.

I have been noticeably psychic since before the age of two, and these abilities have increased over the years. No doubt DNA from my father's side of the family had something to do with this. My patriarchal line had genius level IQs and were outstanding in their fields of engineering, naturopathy, chiropractic, boxing, music, languages and theater. My father, grandfather, and great-grandfather all had a knack for "knowing" what to do and when and were very intuitive. My mother was also very intuitive, but no one had my advanced abilities—likely because they didn't practice utilizing their sixth sense as I did daily. I only mention this because I believe by observing my own family and others that there is a connection with intelligence and telepathic communication with interdimensional intelligences.

And there was *The Voice*. From before the age of two, I occasionally would hear a male voice say something in my ear, or more likely inside my head. It was noticeably clear and always sounded like the same voice. Sometimes it would just be one word, and at other times a sentence. The purpose would be to show me something I should pay attention to, such as a caterpillar's cocoon, or a warning such as "Don't go that way." I didn't pay much attention to the voice, and honestly as a child I just thought it was something that everyone experienced and didn't think much about it until much later in life. I found out later that this voice belongs to an Inter-dimensional being who was assigned to me to help raise my level of consciousness, but more about that later.

At age 11, I had my first experience with a ghost. My sister and I got in our beds one evening and turned out the light. Instantly, the apparition of a man's head appeared at the foot of my bed. The head was talking but I could hear nothing. I asked my sister if she could see it, and she opened her eyes. The screams brought my father upstairs, and he turned on the light. The apparition disappeared. After he calmed down my father asked me what the face looked like. He then got a photo album out and I picked out a man in it that looked just like the face I saw. It was my grandfather who died several years before I was born. After that incident, I learned how to communicate with deceased persons, and by age 20 was doing "readings" for people who wanted to communicate with their loved ones who had crossed over.

At age 14 my mother handed me a book by Edgar Cayce, which she thought would help me understand my abilities. This book changed my life, as it made me aware that there were other people like me. I don't remember the title of the book, but it may have been The Power of Your Mind or The Sleeping Prophet. In any case, the book reassured me that I was not crazy, strange, or out of my mind as the kids at school often said about me. I learned early on to keep my mouth shut when I saw or heard something unusual such as a brightly lit orb floating about the classroom at school, or a stranger standing in a corner watching us that no one else seemed to notice, because others would make fun of me. Much later in life, Caycee began to assist me with readings by recommending certain things, especially castor oil for neck and back pain, and liver or stomach problems with exterior compresses.

My family moved to Springfield, Missouri when I was 14 years old. The minute I walked into our new house, which was built in the late 1900's, I turned to my mother and said, "This place is haunted." A coldness walked through both of us, and she said, "I know." That house would turn out to be the most haunted house I've ever been in, but it was good training for me to learn how to communicate with spirits. It taught me one important lesson – there is no such thing as permanent death of the soul, only death of the physical body.

One evening I went to my upstairs bedroom, and shortly after, a very bright blue-white light entered the room through the closed shade. This was not a new experience; I'd seen that bright light on several occasions starting when I was a baby. I would find out later what it meant, but at the time I had no idea what was going on. Then a scene unfolded in my dressing table mirror with two women dressed in early 1900's clothing and hats. One woman dressed in black with a big black hat on was stabbing the other younger woman dressed in white. The scene faded away. Somehow, I was shown something that happened in the past. In recent years, I did a study of that house, which was very haunted for the book "How to Research a Haunted House," and my co-author, Violet Wisdom, discovered some information which led to our belief that the scene I witnessed in the mirror actually happened. At the time, I simply thought it was the ghosts in the house, but upon reflection, I now believe that the bright blue-white light was from an extraterrestrial craft helping me to access my sixth sense and realize that I could view past events if I so wished. It was likely Val Thor.

When I was in the ninth grade in high school in Springfield, Missouri, my science teacher brought up the subject of Extra-Sensory Perception. He tested the entire class using several different methods. The next day, the teacher pulled me aside and said that he would like to do some more testing with me during lunch or after school using two other students- one to test, and the other to watch to make sure I wasn't cheating. We did this for a couple of weeks, then he told me we were stopping the testing. I asked why and he said "Because you're scaring me. You're too accurate." I was disappointed, as this was a fun exercise and I would miss participating, but it confirmed that I was very psychic.

I started experimenting with Astral Travel at age 15. I had read about how astral travel worked and began to practice this method late at night when the house was quiet. Astral Travel involves getting your physical body to go to sleep, while keeping your consciousness awake, then using the Astral Body to move about, anywhere on or off planet. I was surprised the first time I managed to do this when I floated up towards the ceiling and turned around to see myself asleep on the bed. I was so shocked that I fell into my physical body with a hard slam, which physically hurt and woke me up. It was quite a while until I attempted that again, but soon I became adept at using the Astral body to travel around the world. What is interesting is when you see other people out in their Astral bodies flying around, too, and you just nod to them as they pass by. I routinely visit a monastery in Tibet and recently found out from some of the monks there that I've spent many lifetimes reincarnating at that location. By Astral traveling, I increased the use of my psychic center, which likely improved my connection to inter-dimensional beings.

By my early twenties I was doing psychic readings for individuals several times per week to supplement my income as a professional musician. My readings were so accurate that many people were astounded at what I knew about them, and what happened in the past or in some cases, future events. At that time, I was able to easily tap into sensing or just knowing things, seeing pictures, and occasionally hearing words or conversations using my sixth sense. I also saw with my mind's eye a person's spirit guides, which were most often loved ones who had crossed over, and occasionally other types of spirits that I could not yet identify.

What I didn't know at that time was that these abilities would increase exponentially in the future and that I would later have direct contact with the entity who had been communicating with me since early childhood.

After a while, I started getting calls from private investigators and law enforcement to assist with unsolved crimes. The deal I made was that they would never use my name in any report or tell their superiors how they got the information. This way, I had some assurance that the criminals involved would never know who found them. I still do this service today at no charge. To date, I've helped solve over 70 large thefts, homicides, and kidnappings.

I discovered out-of-body Remote-Viewing accidentally while working on a missing person case in 1990. A client who is an attorney called me because her 14-year-old niece was missing, and the police were not able to locate her. She had been gone for several hours, and the family was in a panic. I closed my eyes and asked what the girl's name was. My first thought was "How am I going to find this girl?" when suddenly, as I was on the phone with my client, I heard *The Voice* tell me to look down. I did so and saw a thin line going from my solar plexus out into space.

I found myself moving through space following this line in front of me, and then almost instantaneously standing at an intersection. I looked up and saw two street signs. I mentioned the street names and asked the client if she knew where this location was and she said, "Oh my God, that is her bus stop!" I then followed the line and inexplicably moved three houses down the block from that corner and stopped in front of the house. I felt that I was in my physical body in two places at the same time, which I would discover later to be bilocation. I described the house in detail and gave the client the address of the house. She replied, "That is the address where the police went today to ask questions, but they found nothing!"

Honestly, I was shocked at how this was going. I'd never done anything like this and didn't understand how it was working, but it was as if I were standing in the street, looking at this house in person. I then saw the line go up above the house and to the back yard over a shed. I floated up, following the line, and looked through the shed from the top. There was a young teenage girl lying on the floor with her hands and ankles tied and duct tape over her mouth. She was asleep—presumably drugged.

I then moved over the house– this time with no line to follow (I just wanted to see who was in the house). I saw a woman with long blonde stringy hair sitting sideways on a couch, a big man with tattoos sitting in an easy chair, and a man sitting on another couch or chair on the side where I was floating. I was hovering above the three people, but they were unaware of my presence. There was a wood coffee table in the center of the room with drug paraphernalia on it. All three had a can of beer in their hands. These people looked very rough. Somehow, I knew that they were holding the girl for someone else, and that she would be shipped out in the morning to a human trafficking group. They were going to get a lot of money for the girl. I could hear them talking about it while I was there. I then brought my consciousness back to my body.

I told my client that the woman was the one responsible for this. My client responded saying that a woman who lived near the bus stop had made friends with her niece over some time, and that is why they thought to visit the house where she lived first. They thought it strange for a 30-something aged woman to befriend a teenager, and that was their only lead.

I told my client that she needed to get back there immediately, and there was no time to waste. She called the detective who was working on the case and told him she was going back by herself. He said he didn't want her to go alone, so he went with her. She called me later that night and said they found the girl and she was okay. The three people were arrested and taken to jail.

This is just one example of how I discovered an ability accidentally in an instance where I needed to be able to accomplish something. After that experience, I was able to repeat the method I discovered at will. I eventually concluded that I was using my etheric body to leave my physical body via the line that left my solar plexus area, and that I could see events in real time. I was using that body's eyes and ears to see and hear, and this was somehow transferred back to my physical body.

Later, as I worked on a case of a missing person for a Kansas agency, I had to go back in time to view past events. This was a murder case, and I was able to locate the site of the murder, and the site of the burial by looking at both the past and the present. The agency sent a helicopter out and found the body in a well, just where I said it would be. I didn't like working on that case but took comfort in the fact that the family could have some closure.

This was just the beginning though; I was not prepared for the wild ride that would occur over the next thirty years. It seems that some type of intelligence was guiding me in certain directions and assisting me along the way without giving me the complete picture, or in fact any idea as to what would happen next or what fantastic experiences I would have later in life.

Many experiences I had I did not find in books, however, and I have had to deal with them on my own. For example, I had been observing tiny bright white lights moving amazingly fast around trees for many years when, while watching them one evening at age 50, I thought, "I wonder what those are?" To my surprise, the male voice that often spoke to me said, "Why don't you ask them?" The thought never occurred to me to try that, but I sent a telepathic message out to the lights, and I heard a response: "We are tree sprites. We stay with a tree for its lifetime, tell it where to grow new shoots and leaves, and when to go dormant." I was amazed, and now I had an answer to that long-standing question. After communicating with the sprites, I also had an answer to another question about something I had observed– I see the energy field of stems and leaves before they grow out, and I know how big the tree will get over the next growing season because the energy field, built by the sprites, exists first.
I found a pattern over the years that involves getting a new or increased ability at or near my birthday. Sometimes it is a small thing – other times it is something incredible.

Introduction to Val Thor

In December of 1985, when I was just 29 years old, an event occurred that was one of those life-changing moments. Prior to this, I had a passing interest in extraterrestrial life and Unidentified Flying Objects. The furthest I had gone in investigating this phenomenon was to join a UFO study group in Seattle, Washington, where I lived for 13 months. At that time, I didn't know why I was interested in UFOs. I just had an interest in them and wanted more information.

On my December 10 birthday I visited my massage therapist. I brought my younger brother and my two children with me because we were going to the Seattle Fish Market right after my visit. As we were putting on our coats to leave the massage therapist stopped me and asked if I would like to try a light experiment. His voice didn't sound like his voice, and I knew that something out of the ordinary was happening, so I played along and said I'd do the experiment. He then asked me to look at a green light, then a red light, and back and forth until he said, "Now what do you see?" I looked up at him and was amazed to see only a skeleton standing there talking to me. I replied in shock, "I see your skeleton." (*Note: I've since discovered that a few other people have had similar activation experiences.*)

Then a face appeared on the wall to my right. The face was a clean-shaven man's face wearing a helmet that came to a point at the top of his forehead and down the sides in front of the ears. This man said "I am Thor, and now you have X-Ray Vision. I will be in contact with you soon and we will be working together." Then the head disappeared, and I looked back to see my massage therapist standing there as normal, but I don't believe he knew what had just happened. I believe that the entity called Thor spoke through the therapist and took over the conversation. I then asked my brother and two children what they saw, and the all said they didn't see or hear anything. Apparently, this experience was just for me. It would take some time before I understood what had happened, but from that point on I could see through the human body to not only see the skeleton, but also organs, veins, glands, blood, etc. and everything is clear and in color. I began to do health scan readings and accurately point out problem areas in people's bodies, and still do that today. It not only amazes the audience when I do presentations for large groups, but me, too. I realize of course, that if it weren't for Thor, I would not have this ability.

A couple of days after that encounter, my brother asked me to do a reading for him. I closed my eyes, but then I heard a voice say, "open your eyes." I did so, and to my amazement, there was Thor again on the wall, only this time his head was much larger. He said "I will be working with you on projects in the future. My ship will be visible to you at your next UFO meeting in Seattle." And that was it. He was gone once again and didn't hang around for questions. I told my brother what happened and said that I thought that Thor must be an extraterrestrial, although I couldn't be certain if he was or not.

We went to the next UFO meeting, and I positioned myself close to the picture window in the house so I could keep an eye out for the ship that was to appear. Halfway through the meeting I felt a presence, and I noticed that one of the other women did as well. Then I was amazed to see a sliver saucer-shaped craft float down and almost touch the ground of the street and both yards on either side of the street. The craft was completely silent, although we all felt a vibration. It had different colored lights on it. However, no one could see the craft except me. I was frustrated that Thor only allowed me to see it and I had no witnesses, not even my brother- but for me it was confirmation that this entity who called himself Thor was real.

From that time until 2006, I had contact with Thor only audibly, through telepathic communication. I came to realize that it was Thor's voice all along, ever since my early childhood, that I heard speaking to me. Eventually, I came to the realization that if I wanted to ask a question of Thor, he was available for me to make the initial contact. Thor told me that he wanted me to author a book about my own experiences and start Un-X News Magazine, and later the Un-X Network digital broadcasting, and write a book about him.

In 2006, things would take a dramatic turn with Thor, and I got more information about who he is during a very strange incident. I left a store I had on the Independence Square to drive home one evening when I noticed heavy cloud cover and strange lightening in the clouds, but no thunder. As I drove, I saw a strange beam of light appear through the cloud to my right at a 30-degree angle upwards. I thought to myself "I wonder what the heck that is?" when to my amazement, Thor's voice said, "I am Commander Thor, and I am on a ship." At the same time, I had a vision of an exceptionally large saucer-shaped craft in the clouds and many others like it at much higher altitude around the entire planet in a grid-line pattern. Information came in telepathically, not in sentences, but more of an overall picture. Thor and his fleet are guardians of the planet Earth and one of their missions is to protect us from negative forces who wish to use the Earth for resources. Another mission is to help raise consciousness of mankind, which they do through certain individuals. The beam of light then went out and Thor was gone. I had a better understanding of who this entity is but would find out more surprising information later.

In practicing using my psychic center daily, and meditating every evening, along with doing private readings for individuals on a regular basis, this has kept my pineal gland and psychic center active. But I fully realize that these talents were not developed on my own – they were gifted or activated by higher intelligences such as Thor and others.

Several incidents occurred which lead me to the conclusion that there are always higher intelligences living right next to us in other dimensions, and that these beings have an interest in helping mankind develop in a positive manner.

Second Contact:

My second fully conscious connection with non-human beings occurred in the year 2000. I was in a dead sleep when at exactly 3:00 O'clock a.m. I heard my grandfather's voice very clearly say in his Brooklyn/Italian accent "Margie get up." I was instantly awake. The voice had come from my doorway which was 15 feet away. *Note: this location is where many entities have entered my house.* My grandfather died in 1972, and I had not communicated with him since, even though I had tried on many occasions. I then started to lay back down when I heard the same voice say the exact same thing, only this time it was more impatient. I sat up. Then my grandfather instructed me to go downstairs to my secretary's desk and sit at the computer. I went into a light trance-state and listened while he spoke, then typed what he said. I had my eyes closed the entire time and at when he finished speaking, I had three full pages of information. Up to that point, I'd never been successful at automatic writing, but discovered that it could be done by typing.

Next, without my grandfather's assistance, other beings made themselves known. A seven-foot-tall green Mantis being appeared one day to my amazement and did a past-life regression to show me a previous life about 1,100 years ago. This was around Montana or South Dakota. He first showed me myself at age 5 in that lifetime where there was a war between two tribes of people. A man on a horse was present but did not participate in the fighting. Both of my parents were killed, but this man spared my life after looking into my eyes and seeing something he thought was special. The Native American man was a shaman, and he was an incarnation of White Cloud. Next, he showed me several scenes of a large village with hundreds of Teepees and fires. This was next to a large clear stream and in a valley with mountains not far away. The shaman took me in and taught me everything he knew about healing. When he died, I took over and became known as Medicine Woman.

At one point in that incarnation, I needed to go on a vision quest and walked up one of the mountains. While standing on top of the mountain on a flat area a silver disk-shaped craft appeared above me and to my right. Then the Mantis being stood to my left. The Mantis then said that he has been with me through many incarnations and that he was assigned to me as a guide. To my surprise, we then entered the craft and traveled very quickly to another solar system. As we approached a planet, he said "This is the Crystal Planet." The planet had a smooth, shiny outer covering of crystal. I realized at this point that the memory of this place must have stuck with me for generations of incarnations since I've always been fascinated by crystals, have dug my own crystals, and even had business cards made up that said "Crystal Planet" on them. I fully intended to sell crystals but never did. I kept them all or gave some away.

Next, we entered the planet through a large opening, and I found myself in the center of a room with a clear wall around it. The Mantis was in the room with me. There were 12 other Mantis' standing on the other side of the wall, and the Mantis explained that they could not be in the same atmosphere as I was. . They each had a pen-like object in their hands, and they aimed these at me. Each object emitted a different colored beam of light that looked like a laser. It implanted small "seed" in each one of my light bodies, which the Mantis explained would activate at a certain time in the future. I was given to understand that the light bodies stay with a person throughout eternity, and that only the physical body dies. The Mantis said that this is one way that the guardians assure that people become aware and awake. It was at this point I realized that there is much more going on in our reality than most people believe. I intuitively knew that there are multiple dimensions and multiple types of beings with various levels of understanding and that mankind is on one of the lowest rungs of the ladder.

The Sixth Sense

The so-called sixth sense is a combination of multiple subtle senses. When the psychic center is activated and used often, many new senses will become apparent.

Clairsentience is the first intuitive talent that people report having. It means that you sense something without the benefit of seeing or hearing it. You just "know" something. This is the first stage of psychic awareness and is usually called "intuition" by laypersons. Intuition is associated with persons who know that Aunt Betty is going to call on the phone in a few minutes, or that someone is talking about them at that moment. But it can go beyond that, to include investigative work with other people.

The intuitive learns to trust what he sees or hears or feels that is beyond the normal five senses. This is difficult at first, because it may not seem "real." The reality is, however, that intuition *is* real, and it becomes clearer as you practice using these talents. For instance, if you stick your hand in a freezer, you will instantly feel physically cold, but if a spirit passes by you may feel a coldness, or sense a coldness, which in most cases will be much less obvious than the freezer cold. This is one example of the subtleness of the experience.

Clairaudience is simply hearing the so-called "other side," or in some cases guides or even advanced beings not in a human body. Usually, this means hearing the voice of a spirit talking, but it can also be the sounds of a plane, traffic, or other things that might be an important part of a message. If a person is trying to communicate that he was killed by a train I may hear a train whistle blowing, or if they were murdered by a gun, I'll often hear the gunshot.

Many people experience clairaudience when they hear their name called or hear someone knocking on their door and find that no one is there. This is your own higher self which is trying to get you to pay attention to something. I've heard people say that they heard their name yelled out and it woke them up in the middle of the night, and they found that their house was on fire or something else was going on that needed attention.

I experienced this when I was 16 years old. My mother was driving from Springfield to Kansas

City on Highway 13 and the road suddenly became icy. This was at a time when no one wore seatbelts. I heard my grandfather's voice say, "put your seatbelts on now." and my mother and I did so, but my brother inexplicably got down on the floor in the back. A couple of minutes later, my mother lost control of the vehicle, and we spun on the ice, then slammed into a guard rail head on. If we hadn't had our seatbelts on, we would surely have gone through the windshield. I recommend that everyone adhere to any messages they receive in this manner!

Clairvoyance is similar, in that you use your third eye, rather than your physical eyes for most of this work. I close my eyes when contacting spirits because I see better that way using my third eye.

Using the third eye is quite different. What you need to see or what is important will appear to you very clearly, in color, with enough light to get a good picture. For me, it is like watching a movie, but for some they may just get glimpses of scenes. If I want to go back in time to watch a particular event, I'll often see the location with all the surroundings, the people present, what they were wearing, etc. However, this method is not 100% accurate when it comes to timing of the event.

Time is the absolute hardest thing to determine, and any clairvoyant will tell you the same thing. If you ask to go to a particular event, you'll have more luck than if you ask what the date was. This must have something to do with our perception of time. To get around this problem, I'll often look around the room I'm visiting for a newspaper or magazine with a date on it and sometimes I'll get lucky.

Synesthesia is the ability to see music and hear color, at least in my case. Other synesthedes have different abilities. I'm not sure if this is an actual Sixth Sense, but it is unusual and in speaking with others, I've found that they are also sensitives and psychics. I discovered this ability at age 17 when playing the bass in an orchestra and suddenly moved outside my body (bi-located?) and watched as streams of color emitted from each instrument and went to the audience. A voice (Thor) said, "You are watching healing take place." Since that time, I have had the ability to see the vibratory rate of music and color, and also people and living things.

Telekinesis is the ability to move objects or affect things in some manner by using the mind. I discovered that I have telekinetic abilities if I am angry or stressed about something. For some reason, the intense emotion of anger or stress allows me to lock or unlock doors, turn lights off or on, or unfortunately kill or injure living things. I found out the hard way when I accidentally killed a fish and later made two people ill. So, I do have to keep my emotions under control. It seems that both good and bad things come with certain abilities. And no, I do not bend spoons – I just don't see the point.

Bi-Location is a term used to describe being in two places at the same time. I discovered by accident that I have this ability while remote-viewing Isla Mujeres in Mexico in 2018. Two of my friends happened to be there at the exact time I was remote-viewing and saw me as a solid person. They called just to make sure it was me. Since then, I have bilocated at will. I'm convinced that this is done not with a subtle body, but by actually creating a second, solid 3-dimensional body. As with most things, I've discovered new abilities by accident.

Remote Viewing is the ability to send your consciousness to a location and time to view and hear events, or to look inside objects, bodies, or buildings. It does not matter when an event occurred because your consciousness travels outside space and time. I found out while remote viewing an event where a person was taken by extraterrestrials, that the ETs could see me when I popped into their craft, leaving no doubt that I was there.

So far, I've discussed my own experiences using the sixth sense and how this developed over time. Now I'm going to attempt to explain the sixth sense and how communication works with higher-dimensional beings. None of this, at least to my knowledge, can be proven scientifically, however, that does not mean that it doesn't exist. It simply means that science, for the most part, has not developed equipment that can detect the fine energies involved, apart from some newer equipment used in paranormal investigations.

It is important to understand the physiology of the part of the brain that is responsible for the sixth sense, which is where remote viewing begins. The pineal gland is one of the most mysterious and misunderstood glands of the human body. Some believe that it has many mystical properties. Science defines the pineal gland as a pinecone shaped gland located at the middle of the brain. It produces melatonin, helps regulate sleep patterns, and influences sexual development. However, mainstream science, unlike some religions, is missing some of the most notable features of the pineal gland (hence the pine-cone shaped object depicted in many religions across the world).

A few years ago, I was meditating and decided that there must be a scientific reason for psychic abilities and remote viewing and asked my guides to give me an answer. I knew that there simply had to be a scientific and physiological explanation for sixth-sense abilities. I immediately saw myself sitting in a chair, and a line going directly into my head and brain, pointing to the pineal gland. I had to look this up online to make sure that I was looking at where I thought this glad was located. It was correct, and I found a lot of useful information about not only the pineal gland but the entire psychic center. My guides were right.

Besides regulating sleep patterns and sexual development, the pineal gland acts like an inner portal that connects us to other dimensions, such as the dream and spiritual planes. Some researchers believe that the pineal gland can connect us to other dimensions. I agree that this is the case, as I've also been able to see and communicate with beings from different dimensions.

People have been using their psi center or "third eye" for millennia. It is believed that the ancient Egyptians use the symbology of the Eye of Horus to represent the part of the brain that is responsible for psychic awareness.

Seers, mystics, wizards, and enlightened ones, all utilize the psi center to various degrees. This is likely the center of power used to do telekinesis, telepathy, bi-location, and remote-viewing.

The pineal gland is shaped like a pinecone and taken from an article written by Richard Cassaro on this subject: *"The pinecone symbol is one of the most mysterious emblems found in ancient and modern art and architecture. Few scholars realize it, but the pinecone alludes to the highest degree of spiritual illumination possible. This was recognized by various ancient cultures, and the symbol can be found in the ruins of the Indonesians, Babylonians, Egyptians, Greeks, Romans, and Christians, to name a few. It also appears in the drawings of esoteric traditions like Freemasonry, Theosophy, Gnosticism, and esoteric Christianity. The pinecone held the same meaning for all: It symbolized a secret vestigial organ, the "pineal gland" or "Third Eye," that we all possess."*

The Third Eye

The Third Eye, Eye of Horus, or Psychic Center is the connection from the physical body to the subtle bodies, which in turn communicate with inter-dimensional intelligences. Therefore, the only way for an individual to experience seeing or hearing beings in higher dimensions is to activate their PSI center. Obviously, this cannot be done for many because they simply are not yet ready to understand or experience using their PSI center because they may believe that it is unscientific, or borderline insanity to believe it could work. But many may not be able to activate their PSI center because the pineal gland is calcified. Calcification occurs when Sodium Fluoride is consumed over a period of time.

During the late 1990's in England, a scientist by the name of Jennifer Luke undertook the first study the effects of sodium fluoride on the pineal gland. She determined that the pineal gland, located in the middle of the brain, was a target for fluoride. The pineal gland simply absorbed more fluoride than any other physical matter in the body, even bones.

Because of the pineal gland's importance to the endocrine system, her conclusions were a breakthrough. Her study provided the missing link to a lot of physiological damage from sodium fluoride that had been hypothesized but not positively connected. A veritable root source for the chain reaction of blocked endocrine activity had been isolated.

One could theorize that the addition of Fluoride to our drinking water, toothpaste, and other products may have been purposeful to keep the population from activating their PSI center.

Frequent exposure to outdoor sunshine, at 20 minutes or so at a time, will help stimulate a fluoride calcified pineal gland. This is more important than most realize because the pineal gland affects so many enzyme and endocrine activities, including melatonin production and the production of "Dimethyltryptamine" DMT which gets secreted when you born, when you dream, and when you die.

By keeping my Pineal gland and PSI center active, I've been able to see and communicate with higher intelligences from different dimensions. This includes contact with multiple non-human intelligences which have appeared physically solid or transparent both inside and outside my house; telepathic communication with extra-terrestrial and inter-dimensional beings; remote-viewing of inter-dimensional and third dimensional locations; and seeing the "Fast-Movers," which are inter-dimensional craft that move so fast most people can't see them with the naked eye.

The Fast-Movers have been captured on video, but I see them during the day or night, and by simply wanting time to slow down, I can get a good look at them. I have come to understand that what is happening is that I am raising my vibration to a higher dimension, so things appear to move more slowly. The Fast Movers appear as typical saucer type craft and non-typical craft of varied sizes and shapes. I recently found out that these high-speed objects have been observed since at least the early 1990s by our military and individuals.

Help from Inter-dimensional beings

Solid and transparent beings have appeared inside my house on numerous occasions. The first such incident occurred several years ago when I had some severe pain in my feet for which there was no explanation. It got to the point that I could no longer walk and had to crawl on the floor to get around. After a few days of this and with no help from my doctor. I lay down on my bed one evening and sent a telepathic message out to the Universe for help. I then fell asleep, but was rudely awakened later by a sharp pinch, as if a long needle were inserted into my left calf into the bone. I opened my eyes and saw two 4' tall non-human beings fully formed at the foot of my bed. One had his hands out below my feet and the other stood to the left behind the first. Both beings had humanoid bodies and large heads but did not look like the typical Greys. They had normal sized eyes, small noses and ears, and small mouths. I sensed that there was a third entity to my right but did not see it. My first reaction was to scream and throw the covers over my head, and when I finally gained the courage to lower the covers the beings were gone. At least, I no longer saw them. The next day I noticed that the pain in my feet was completely gone. In analyzing this incident, I realized that I asked for help, and help came, and that I was able to see the entities due to my heightened sixth sense and awareness but soon lost that ability once I was in the lower emotion of fearing what I saw. This was a good lesson.

There have been many other encounters which are too numerous to mention in this chapter but suffice it to say that I'm fully convinced that there are multiple inter-dimensional intelligences who will come to our aid if asked, and who are available to assist those of us who wish to raise our vibration and level of consciousness. They literally have answers to just about every subject you can imagine. I'm also convinced that persons such as Nikola Tessla, Albert Einstein, George Frederic Handel, Ludwig Von Beethoven, and others were in contact with higher intelligences, and some even admitted this. These people were helping mankind advance, so perhaps some of the higher-vibration beings noticed and decided to assist them to speed up the process.

In my case, I asked Thor how many people are like me when it comes to communicating with inter-dimensional intelligences and other abilities, and he said one in 100,000. Thor said that the councils (groups of beings from higher dimensions) choose certain people to work with, which in turn are charged with helping to raise the vibration and consciousness of others. This is the reason why I wrote *The Remote-Viewing Workbook* and developed the Remote Viewing Workshop, and why I do seminars and speak at conferences.

All humans must have these abilities, but most need assistance getting to the point of awareness, then they have the choice to act on their knowledge or not. By choosing to act on it, the person begins the path to enlightenment which eventually leads to direct contact with higher intelligences in different dimensions.

A Quantum Communication Experience
Why do some people see UFOs and others don't?

This question has been asked many times, and it is one that I often get from experiencers. One case in point that I often think about occurred in the 1990's in downtown Kansas City. A woman said that she was on her lunch break with three other woman, and as they walked down the street she noticed a very large disk-shaped UFO hovering over the buildings to her right across the street. The others did not notice it – and she asked them to look above the building and said, "Do you see that?" To which the reply was "See what?" from all three. This was very confusing to the witness, and she asked me if she was crazy. I told her she wasn't crazy and that this sort of thing happens to people who have raised their level of consciousness.

I had been investigating portals when during meditation one day I asked, "How is it that I see UFOs?" The answer was shown to me in a vision of me standing in my driveway during a real past event. I watched from 80 feet up from the ground and the same distance away, as a bright flash of light appeared all around me (the me standing on the ground) as I stared up at the northern sky. At the time of the actual event, I did not see the bright flash, only a UFO which appeared out of nowhere and stayed in that spot for several seconds. Then Thor's voice spoke, "You just shifted into a higher dimension where the craft is and that is why you can see it now. The flash is the opening of a portal, which you created by shifting your consciousness."

That explained a lot. Not only does it explain the flashes of light some people see before seeing UFOs, or for that matter, inter-dimensional beings, but it is the reason why, by shifting our own consciousness, we can move between dimensions. And we can create our own portals in order to do so!

Conclusion

Seeing UFOs and extraterrestrials requires a shift in consciousness to a higher vibration and dimension. I see vibration when music is played, and around all things. The higher the frequency, the higher the dimension. Think of dimensions as a keyboard that has multiple octaves – when you get to the top or bottom of the keyboard and keep going, eventually the notes cannot be heard in our Third Dimension, but they can be heard in higher dimensions. The same goes for the visible light spectrum, which is a very narrow band as far as what the normal human eye can see, but if special cameras or equipment are used, the higher and lower spectrums become visible. This is why some people hear and see outside the normal sound and light spectrums – simply because they have raised their level of consciousness.

As I mentioned in listing some of my abilities before, these are available to everyone and with practice, they can be fully developed. By practicing ESP, Remote Viewing, Meditation, or meditative tasks such as coloring, painting, music, etc., a person can activate their PSI center, which in turn allows them to see interdimensional beings and craft, and eventually have direct telepathic communication.

This is available to everyone – that is, anyone with an open mind who wishes to pursue it. I highly recommend trying Quantum Communication for yourself as it is a rich and rewarding experience. We can all learn from our Star Families, who are more than willing to share information. Their goal is to raise the consciousness of planet Earth.

CE5 and CE6 Explained

CE5, or Close Encounters of the Fifth Kind, is a term created by J. Allen Hynek and is now used as a distinctive approach in which people initiate and build contact with interstellar beings and other extraterrestrial intelligence through conscious practice. Meditation and non-verbal communication techniques sit at the center of the method.

The story of CE5 often points back to Dr. Steven Greer, a leading figure in ufology and the founder of the Center for the Study of Extraterrestrial Intelligence (CSETI). Greer's long engagement with UFOs and extraterrestrial phenomena pushed him toward a more proactive idea: peaceful communication and cooperation with potential extraterrestrial civilizations. Through lectures, books, and films like "Unacknowledged," he helped bring CE5 into wider view among UFO enthusiasts and researchers.

The framework for "close encounters" has older roots. In his 1972 book, "The UFO Experience: A Scientific Inquiry," Dr. J. Allen Hynek introduced a classification system that included Close Encounters of the First, Second, and Third Kind. Later, C. D. B. Bryan and Dr. Greer argued that two more categories were needed.

Earlier labels leaned on passive observation (CE1, CE2, and CE3) or abduction scenarios (CE4). CE5, by contrast, reframed the encounter as human-initiated contact, and the phrase "Fifth Kind" highlights that proactive, deliberate communication attempt with extraterrestrial entities.

Groups are now hosting CE5 events worldwide with great success in contact with extraterrestrials and inter-dimensional beings.

Definitions:

CE-1 – Close Encounter of the First Kind (Observation at Close Range)

Definition: Visual observation of a UAP at close range (often ≤ 500 ft / 150m) with appreciable angular size and detail; no claimed physical effects.
Use in fieldwork: Baseline for collecting angular size, estimated distance, duration, azimuth/elevation, and sensor corroboration (cameras, radar).
Notes: Many CE-1 reports today involve misperceptions of satellites (e.g., Starlink trains), drones, and re-entries, requiring robust correlation with astronomical/space-traffic data. Space

CE-2 – Close Encounter of the Second Kind (Physical Effects)

Definition: CE-1 plus one or more physical effects: EM interference with vehicles/electronics, animal reactions, physiological sensations (heat, tingling), or trace evidence (soil compaction, scorch marks, residues).
Use: Demands forensic protocols, chain-of-custody for samples, blind lab analyses, site controls, and careful confound handling (soil moisture, vehicle tracks).

CE-3 – Close Encounter of the Third Kind (Entities Observed)

Definition: CE-1/CE-2 plus observation of an entity (humanoid/robot/other) associated with the UAP.
Use: Requires composite sketches, independent witness interviews, and tracking of Bloecher subtypes (e.g., "Aboard," "Close," "Direct"). Center for UFO Studies

CE-4 – Close Encounter of the Fourth Kind (Abduction / Transformational)

Definition (common use): Abduction or forced transport by a UAP or its occupants, often with missing time, medical procedures, and long-term aftereffects; some researchers broaden to transformational reality shifts.
Use: Demands clinical-grade interviewing, sleep/parasomnia screening, corroboration attempts (co-witnesses, environmental traces), and avoidance of suggestive techniques in testimony collection.

CE-5 – Close Encounter of the Fifth Kind (Human-Initiated Contact)

Definition (contemporary popular use): Deliberate, protocol-driven attempts by humans to attract and communicate with non-human intelligences: group meditation, coherent thought sequencing, geolocated vectoring, structured light (including laser pointers in some groups), and audio signaling. Sessions are often documented with night-vision, astronomical apps, audio recorders, and synchronized logging. Dr. Steven Greer

Critical safety/legal note: In the United States it is a federal crime to aim a laser pointer at an aircraft or its flight path (18 U.S.C. §39A); the FAA treats unauthorized laser illumination of aircraft as an in-flight emergency, with formal reporting procedures and safety guidance. Any CE-5 protocol must rigorously avoid sky-directed lasers and manage line-of-sight risks near air traffic. Federal Aviation Administration

CE5 Protocol (my method)

Step 1: I usually conduct a class in order to familiarize everyone participating with the methods used and why they must be done in a certain order. I explain that I am the only one allowed to use a laser pointer due to FAA regulations, and not to talk during the meditation session.

Step 2: Have everyone set up their lawn chairs, or sleeping bags outside. Tell them to bring blankets to keep warm and comfortable.

Step 3: The meditation. I have great success with a drumming circle, either having people bring their own drums or playing drumming music for about 20-30 minutes. I take people through a relaxing guided meditation first with the intention to see our space brothers. This thought is sent out telepathically. This process puts everyone in a meditative state and raises their vibration.

Step 5: Watch for UAPs and ask people to point in the direction where they see something and speak quietly. Have people film what they are seeing. Make people aware that ETs could appear right next to them and to stay calm. I usually know where they will appear and point my laser in that direction in advance of a sighting.

Step 6: You may need to do another meditation session, sending the thought out that you would like to have contact with benevolent non-human beings one or two more times in order to keep people in a high state of consciousness. Three to four hours is the time limit for the contact session because people will get very tired.

Step 7: Compare notes and video – have people share on a designated social media site or ask them to send you their photos/videos for analysis. There may be more on film than what was seen with the naked eye – especially fast movers.

CE6 Explained

This term applies when scientific equipment is used during a CE5 session to determine changes in temperature, barometric pressure, electromagnetic fields, and more. One device I always use is a spectrum analyzer since 1.6 GHz indicates a portal opening, and it often is an indicator when UFOs are in the area. Some of these devices are monitored by a central station, and readings are recorded. The participants record the timing of events, such as UFO sightings, and compare the results with the data. The use of cameras with time stamps is particularly useful.

About the Authors

Wayne Lawrence is a Field Investigator with the Illinois Chapter of MUFON. He enjoys assisting MUFON investigators by examining video evidence and providing an analysis of their video.

In July 2026, Wayne was asked to serve on the MUFON Photo Analysis Team. Previously he was Chief Investigator with the Missouri Chapter of the Mutual UFO Network. Lawrence noticed objects moving in and out of Blue Springs Lake, MO, in 2014, and paid particular attention to the activity in that area. Many of them are very fast-moving craft. He has made numerous video lecture presentations that exhibit images and data from those cases that he has investigated.

Wayne has been a frequent presenter at local and state MUFON chapter meetings and paranormal conferences in Missouri and with other MUFON chapters in the US. His talks included topics such as Buildings on the Moon, Anomalies on Mars, Solar Visitors, "*The Fast Movers*," Highlights of his own sightings, How to Investigate Sightings Using Math Formulas, High Strangeness, Cubes, UFOs and Nukes, and more.

Wayne had a case that was included in "*UFO Cases of Interest, 2019 Edition*". A number of his investigated cases were highlighted in the national publication "*The MUFON Journal*" over the last several years. His article in 2023 was entitled "*When New Features are Not Helpful*" which he co-wrote with two other investigators about night mode in cellphones. His previous book co-written with Margie Kay and Bill Spicer is entitled "*The Fast Movers*." The latest book is entitled "*Earth's Unknown Inhabitants*" is expected to be published in late 2026.

Lawrence grew up in French Polynesia where he learned three languages. He is currently fluent in five languages. He travelled extensively around the world and for over twenty years was Director of Language Translations for an international denomination.

He retired at the end of 2020, and lives with his wife in Naperville, Illinois. He has two married children and three grandkids.

Bill Spicer is a near death experiencer (NDE), completed a work career as an engineer on multiple aircraft and who's past time includes amateur astronomy using IR Night Vision techniques. As a UFO/UAP researcher, he uses a fascinating technique of capturing photos and videos of UFO/UAP Craft with digital cameras since 2009, that he has called Quantum UFO Observation Technique. This is used to see "The Fast Movers," UFO/UAP Craft. He is the author of "Introduction to aircraft Interiors" and "Beyond Watching- A Journal of Experiences Understanding UFOs."

May of Bill's photographs of Fast Movers can be viewed on The Fast Movers Facebook Group and available to the public.

Bill is the co-author of The Fast Movers, which was published in 2023. He has spoken at several meetings and conferences, demonstrating his techniques and training others to get the same results.

Lily Nova is an ET contactee, world-renowned UFO and space photographer, cosmic channel, and best-selling author of "Contact." Her story and UFO footage have been featured on Ancient Aliens, The Travel Channel, and mainstream media all over the world, including LAD Bible and The Mirror.

Through her contact experiences, Lily has activated extra-sensory perception and regularly shares cosmic insights on her YouTube and Patreon channels. She is also a public speaker and hosts spiritual retreats in high energy locations and paranormal hot spots like Sedona, Arizona and Mount Shasta, California.

Website: LilyNovaContact.com

Margie Kay is a veteran 40-year paranormal investigator. She is the Owner and CEO of the Unexplained Network and Un-X Media publishing company, serves as MUFON Central Region Director, State Director for Missouri MUFON, and as the Director of the OZ Interdimensional Communication Institute. She is a licensed private investigator in Missouri and owns and operates a fire investigation company. Margie has completed over 780 UFO investigations for MUFON and over 1,900 paranormal and UFO investigations on her own.

Kay is a self-taught psychic medium and remote viewer and has helped to solve over seventy homicides, theft, and missing person cases for law enforcement, private investigators, and individuals in the U.S., South America, and Europe. She also assists other investigators with unsolved UFO and paranormal cases.

Margie is an accomplished classical string bass player and jazz/fusion electric bass player and worked as a professional musician for 20 years. She launched the KUNX Digital Broadcasting Network on October 31, 2021. There are multiple paranormal programs on the network.

Margie is the author of The Kansas City UFO Flaps; The Fast Movers: Evidence of High-Speed UFOs; Haunted Independence Missouri; Gateway to the Dead: A Ghost Hunter's Field Guide; The Remote Viewing Workbook; Winged Aliens; The Masters Dowser's Chart Book; A Sonoma County Phenomenon; How to Research a Haunted House, and more. Some of Margie's experiences are chronicled in the books "*Paranormal Missouri: Show Me Your Monsters*." and "*What Lurks Beyond: The Paranormal in Your Backyard*" by Jason Offutt, and in "UFOs in Missouri" by Lee Prosser. Kay also contributed a chapter to the book "A Greater Reality" by Rey Hernandez at his request.

Margie Kay has hosted radio shows since 2003. She has interviewed dozens of paranormal and UFO researchers and authors, including Kathleen Marden, Stanton Friedman, Edwin Mitchell, Craig Campobasso, and many more. Margie's Un-X News Radio/Podcast on her YouTube Channel is broadcast weekly. She has appeared on numerous radio shows and on the pilot TV show Strange; on Hangar 1 UFO Hot Spot; Ancient Aliens, UFO Witness (The Cube) with Ben Hansen, and news programs on local channels and CNN.

Margie has written over 250 articles for magazines and news sites. Her articles have appeared in the MUFON Journal, Un-X News Magazine, the Midwest MUFON News, Examiner.com, The Un-X Blog, and more.

Kay has spoken at over two hundred conferences and meetings nationwide and she created multiple on-demand workshops about Remote Viewing, Chakra Balancing, Healing with Music, Psychic Protection, Esoteric Knowledge, Communicating with Extraterrestrials and more. Her students have had remarkable success using her unique methods.

Websites: www.margiekay.com
www.unxnetwork.com www.unxmedia.com
Blog: un-xnews.blogspot.com
YouTube Channel: UnX News Radio/Podcast
youtube.com/channel/UCUJYHq_4FmsK_6ng5y8yeDQ

Participate in CE5, CE6 and Quantum Communication Events

Learn from the experts and have your own experiences

Events with Lily Nova

Lilly Nova is an ET contactee, world-renowned UFO and space photographer, cosmic channel, and best-selling author of "Contact." Her story and UFO footage have been featured on Ancient Aliens, The Travel Channel, and mainstream media all over the world, including LAD Bible and The Mirror. Through her contact experiences, Lily has activated extra-sensory perception and regularly shares cosmic insights on her YouTube and Patreon channels. She is also a public speaker and hosts spiritual retreats in high energy locations and paranormal hot spots like Sedona, Arizona and Mount Shasta, California.
Website: LilyNovaContact.com

Events with Margie Kay

Margie Kay is an ET contactee and internationally acclaimed remote viewer, psychic medium, author, and podcast host. She is in communication with Val Thor, a Mantis being, White Cloud, Medicine Woman, and an entire Sasquatch Clan among others. Kay teaches multiple workshops on a variety of subjects on demand and in person, and hosts in-person CE5 and CE6 training and contact events across the U.S. with amazing results. She also does private readings and healing sessions for individuals. Kay is a Certified Sound Healer and Reiki Master, professional musician, and radio and TV personality.
Website: Margiekay.com

Suggested Class: ***How to Contact Extraterrestrials***, is an on-demand workshop by Margie Kay. Visit www.margiekay.com for more information. A meditation workshop is forthcoming.

Books by Un-X Media

Un-X Media publishes fiction and non-fiction books about unexplained phenomena, including UFOs, Aliens, Cryptids, Consciousness, Conspiracies, and more.

Non-Fiction Books

Haunted Independence, Missouri 2013 - 2026 by Margie Kay
Gateway to the Dead: A Ghost Hunter's Field Guide 2013 by Margie Kay
The Kansas City UFO Flaps 2017 & 2026 by Margie Kay
Family Secrets 2017 by Jean Walker
The Remote Viewing Workbook 2019 & 2026 by Margie Kay
A Sonoma County Phenomenon 2020 by Margie Kay
The Color Therapy Wall Chart by Margie Kay 1999
Rules for Goddesses by Margie Kay 2000
Doorway to Spirit by Devin Listrom 2020
The Alien Colonization of Earth's Waterways by Debbie Ziegelmeyer 2021
The Master Dowser's Chart Book by Margie Kay 2022 (lulu.com)
Dying to Meet Them: One Woman's Incredible Journey from NDE to UAP by Mindy Tautfest 2023
50th Anniversary of the SE Missouri Ozarks UFO Flap by Debbie Ziegelmeyer and Margie Kay 2023
The Fast Movers by Margie Kay, Bill Spicer, and Wayne Lawrence 2023
The Varginha Incident by Vittorio Pacaccini and Fernanda Pires 2024
Un-X News Magazine 2001-2015 and 2001-2026
How to Research a Haunted House by Margie Kay and Violet Wisdom 2024
Alien Attacks in Brazil by Thiago Luiz Vicchetti 2025
Take a Trip on Haunted Route 66 by Margie Kay 2025
We Got it from THEM! by Dr.Gregory Rogers 2025
THOR: The Extraterrestrial on Earth by Margie Kay 2026
Deepwater UFOs: Alien Residents and Abductors by Debbie Ziegelmeyer 2026
Top UFO Cases in the Midwest by Central Region MUFON 2026
Earth's Unseen Inhabitants by Margie Kay, Bill Spicer, Wayne Lawrence and Lily Nova 2026

Fiction Books

All Monsters are Human Part I by Derrick Smith (horror fiction) 2025
Real Fear by Derrick Smith (horror fiction) 2025
IMPACT: Special Edition by Dr.Gregory Rogers 2025
An Angel in the City by Alice Manette 2026
By Sword and Blood Book One: Oracles of the Dark Mirror! a novel by Gregory A. Rogers 2026
Up in the Sky by Carl Harper, 2026 *And more coming soon*

A Series of Adult Coloring Books
Bookstores—please contact us for wholesale pricing - editor@unxmedia.com

www.ingramcontent.com/pod-product-compliance
Lightning Source LLC
Chambersburg PA
CBHW041321210926
51847CB00002B/15
9781969800078